TOPICS IN PLANT PHYSIOLOGY: 1
SERIES EDITORS: M. BLACK & J. CHAPMAN

LIGHT AND PLANT GROWTH

TITLES OF RELATED INTEREST

Biogeographical processes
I. Simmons

A biologist's advanced mathematics
D. R. Causton

Class experiments in plant physiology
H. Meidner

Comparative plant ecology
J. P. Grime *et al.*

Countryside conservation
B. Green

Hedgerows and verges
W. H. Dowdeswell

Historical plant geography
P. Stott

Introduction to vegetation analysis
D. R. Causton

Introduction to world vegetation
A. S. Collinson

Lipids in plants and microbes
J. L. Harwood & N. J. Russell

Lowson's textbook of botany
E. W. Simon *et al.*

Nature's place
W. Adams

Patterns of life
H. Meilke

Plant breeding systems
A. J. Richards

Plants for arid lands
G. E. Wickens *et al.* (eds)

Process of vegetation change
C. Burrows

Techniques and fieldwork in ecology
G. Williams

LIGHT AND PLANT GROWTH

J. W. Hart

Department of Plant Science
University of Aberdeen

London
UNWIN HYMAN
Boston   Sydney   Wellington

Published by the Academic division of
Unwin Hyman Ltd
15/17 Broadwick Street, London W1V 1FP

Allen & Unwin Inc.,
8 Winchester Place, Winchester, Mass. 01890, USA

Allen & Unwin (Australia) Ltd,
8 Napier Street, North Sydney, NSW 2060, Australia

Allen & Unwin (New Zealand) Ltd in association with the Port
Nicholson Press Ltd,
Compusales Building, 75 Ghuznee Street, Wellington 1, New Zealand

First published in 1988
Second impression 1990

British Library Cataloguing in Publication Data

Hart, J. W.
Light and plant growth. — (Topics in
plant physiology; no. 1).
1. Plants, Effect of light on 2. Growth
(Plants)
I. Title II. Series
581.3'1   QK757
ISBN 0–04–581022–2
ISBN 0–04–581023–0 Pbk

Library of Congress Cataloging in Publication Data

Hart, J. W. (James Watnell), 1940–
Light and plant growth.
(Topics in plant physiology; 1)
Bibliography: p.
Includes index.
1. Plants, Effect of light on. 2. Plants–
Photomorphogenesis. I. Title. II. Series.
QK757.H37 1987 581.3'1   87–14322
ISBN 0–04–581022–2 (alk. paper)
ISBN 0–04–581023–0 (pbk. : alk. paper)

Typeset in 10 on 12 point Palatino by Columns of Reading
and printed in Great Britain by
Biddles of Guildford

To my wife,
Leona

and my teacher,
A. M. M. Berrie

Preface

There are many recent works on the topic of light and plant growth. These have not only been written by experts, but are also, in the main, written for experts (or, at least, for those who already have a fair understanding of the subject). This book has its origins in a six-week course in plant photophysiology, and its aim is to provide an introduction to the subject at an advanced undergraduate level. The imagined audience is simply a student who has asked the questions: In what ways does light affect plant growth, and how does it do it?

The book is limited to aspects of photomorphogenesis. Photosynthesis is only considered where its pigments impinge on photomorphogenic investigations, or where its processes provide illustrative examples of particular interactions between light and biological material. Chapter 1 gives a general account of the various ways in which light affects plant development, and introduces topics which are subsequently covered in greater detail. In all the chapters, are special topic 'boxes', consisting of squared-off sections of text. These are simply devices for presenting explanatory background material, or material that I myself find particularly intriguing.

Literature citation is deliberately minimized, not only because of space limitations but also in an attempt to clarify the textual presentation at undergraduate level, and great use has been made of the reviews in the many recent edited symposia and encyclopaedic volumes. Relevant material for each section is usually cited at the beginning of the appropriate section, sometimes within the text when greater accuracy of attribution is required. At the end of each chapter is a short list of further reading material. From these and the articles cited with the figures, the interested student should be able to find all relevant original material.

I would like to thank Professor P. G. Jarvis, Dr G. B. James, and in particular the late Dr R. E. Neilson for introducing me to the field of light and its technology (errors are, however, of my own making). Special thanks are due to Dr D. D. Sabnis for his continual interest, advice and stimulating discussion.

I am also extremely grateful to my typist, Mrs Betty Smith, for her unfailing willingness and her unflinching approach to my script.

J. W. Hart

Acknowledgements

Plate 1: from Desroche-Noblecourt (1965) *Tutankhamen*. London: Rainbird, reproduced by permission of the Metropolitan Museum of Art, New York.

Figures
1.1 adapted from Salisbury & Ross (1978) *Plant physiology*, 2nd edn. Belmont, California: Wadsworth. Reproduced by permission; 1.2a from Duggar (1911) *Plant physiology*. New York: Macmillan; 1.2b from Pfeffer (1903) *The physiology of plants*. Oxford: Clarendon Press; 1.3 from Detmer (1898) *Practical plant physiology*. New York: Macmillan; 1.4a from Ashby (1961) *Introduction to plant ecology*. London: Macmillan; 1.4b from Schopfer (1984) in *Advanced plant physiology*. M. B. Wilkins (ed.) 380–407. London: Pitman. Reproduced by permission; 1.5 from Darwin (1880) *The movements of plants*. London: John Murray; 2.1 adapted from Calvert & Pitts (1966) *Photochemistry*. New York: John Wiley; 2.6a from Withrow & Withrow (1956) in *Radiation biology*, vol. 3, A. Hollaender (ed.). New York: McGraw-Hill; 2.6b from Withrow & Price (1953) *Plant Physiol.* **28**, 105; 3.1 from Henderson (1970) *Daylight and its spectrum*. Bristol: Adam Hilger; 3.2 from Smith (1982) *Annu. Rev. Plant Physiol.* **33**, 481. Scans (d) and (f) were computed by Smith from data of, respectively, Salisbury and Spence; 3.4 redrawn from Johnson *et al.* (1967) *Science* **155**, 1663. Copyright (1967) by the AAAS; 3.5 from Whatley & Whatley (1980) *Light and plant life*. London: Edward Arnold; 3.6 from Jerlov (1968) *Optical oceanography*. Amsterdam: Elsevier; 4.1 from Presti (1983) in *The biology of photoreception*. SEB Symposia, no. 36, D. Cosens & D. Vince-Prue (eds.) 133–80. Cambridge: Cambridge University Press; 4.2 from Shropshire (1977) in *The science of photobiology*, K. C. Smith (ed.) 281–312. New York: Plenum Press, redrawn from Flint & McAlister (1937); 4.3 from Butler *et al.* (1959) *Proc. Nat. Acad. Sci. (US)* **45**, 1703; 4.4 from Smith & Daniels (1981) *Plant Physiol.* **68**, 443; 4.5a from Hartmann (1967) *Z. Naturf.* **22b**, 1172; 4.5b from Hartmann (1967) Book of abstracts, Europ. Photobiol. Symp., Yugoslavia, as redrawn by Smith (1975) in *Phytochrome and photomorphogenesis*. Maidenhead: McGraw-Hill; 4.6 & 4.7 from Song (1984) in *Advanced plant physiology*, M. B. Wilkins (ed.) 354–79. London: Pitman. Reproduced by permission; 4.8a from Hanke *et al.* (1969) *Planta* **86**, 235; 4.8b from Smith (1982) *Annu. Rev. Plant Physiol.* **33**, 481, as redrawn from Smith & Holmes (1977) *Photochem. Photobiol.* **25**, 547; 4.9 from Kendrick & Spruit (1973) *Plant Physiol.* **52**, 327 as redrawn by Song (1984) in *Advanced plant physiology*, M. B. Wilkins (ed.) 354–79. London: Pitman. Reproduced by permission; 4.11 data of various authors compiled by Presti & Delbruck (1978) *Plant, Cell & Environ.* **1**, 81; 4.12 from Schäfer (1981) in *Plants and the daylight spectrum*, H. Smith (ed.) 461–80. London: Academic Press; 4.13 from Cosgrove (1982) *Plant Sci. Let.* **25**, 305; 4.14c from Hager (1970) *Planta* **91**, 38; 4.14d from

Sun *et al.* (1972) *J. Amer. Chem. Soc.* **94**, 1730; 5.1 from Epel *et al.* (1980) in *Photoreceptors and plant development*, J. A. DeGreef (ed.). Antwerp: Antwerp University Press. Original photographs kindly supplied by Dr B. L. Epel; 5.2 from Racusen (1976) *Planta* **132**, 25; 5.3a from Haupt (1970) *Physiol. Veg.* **8**, 551; .5.3b and 5.3c from Haupt (1969) *Planta* **88**, 183; 5.5 from Oelze-Karow & Mohr (1970) *Z. Naturf.* **25b**, 1282, as redrawn by Smith (1975) in *Phytochrome and photomorphogenesis*. Maidenhead, England: McGraw-Hill; 5.6 from Mohr (1969) in *Introduction to photobiology*, C. P. Swanson (ed.). Englewood Cliffs. NJ: Prentice-Hall; 5.7 both graphs constructed from the data of various authors by Schopfer (1984) in *Advanced plant physiology*, M. B. Wilkins (ed.). London: Pitman. Reproduced by permission; 6.1 from Frankland (1976) in *Light and plant development*, H. Smith (ed.) 477–92. London: Butterworth; 6.2 from Thomson (1951) *Amer. J. Bot.* **38**, 635; 6.3 from Cosgrove (1981) *Plant Physiol.* **67**, 584; 6.5 from Gaba & Black (1983) *Encycl. plant physiol.* NS 16A, W. Shropshire & H. Mohr (eds), 358–400. Berlin: Springer, as redrawn from Blaauw *et al.* (1968) *Planta* **82**, 87; 6.6 from MacDonald *et al.* (1982) *Plant, Cell & Environ.* **5**, 305; 6.7 both graphs from Jabben & Holmes (1983) *Encycl. plant physiol.* NS 16B, 704–22. Berlin: Springer, from data of Beggs *et al.* (1981); 6.8 redrawn from Cosgrove (1981) *Plant Physiol.* **67**, 584; 6.9 from Jabben & Holmes (1983) *Encycl. plant physiol.* NS 16B. Berlin: Springer, from data of Beggs *et al.* (1981); 6.10 from Morgan (1981) in *Plants and the daylight spectrum*, H. Smith (ed.) 205–21. London: Academic Press; 6.13 from Reid & Leech (1980) *Biochemistry and structure of cell organelles*. London: Blackie & Son; 6.16 adapted from Hanson (1917) *Amer. J. Bot.* **4**, 533; 7.1 from Dennison (1979) *Encycl. plant physiol.* NS 7, W. Haupt & M. E. Feinleib (eds) 506–66. Berlin: Springer; 7.2a from Thimann & Curry (1960) in *Comparative biochemistry*, vol. 1, M. Florkin & H. S. Mason (eds.), 243–309. New York: Academic Press; 7.2b from Curry *et al.* (1956) *Physiol. Plant.* **9**, 429; 7.3a from Haupt (1983) *SEB Symposium No. 36*, 423–42. Cambridge: Cambridge University Press; 7.3b from Shropshire (1962) *J. Gen. Physiol.* **45**, 949. Reproduced by permission of the Rockefeller University Press; 7.4 from Dennison (1984) in *Advanced plant physiology*, M. B. Wilkins (ed.) 149–62. London: Pitman, as redrawn from data of Zimmerman & Briggs (1963) *Plant Physiol.* **68**, 248; 7.5 from Blaauw & Blaauw-Jansen (1970) *Acta Bot. Neerl.* **19**, 755; 7.6a and 7.6b from Thimann & Curry (1960) in *Comparative biochemistry*, M. Florkin & H. S. Mason (eds), 243–309. New York: Academic Press; 7.7 from Shuttleworth & Black (1977) *Planta* **135**, 51; 7.8 from Hart & MacDonald (1981) *Plant Sci. Let.* **21**, 151; 7.11 from Briggs *et al.* (1957) *Science* **126**, 210. Copyright (1957) by the AAAs; 7.12 based on data from Hart *et al.* (1982) *Plant, Cell & Environ.* **5**, 361; 8.1a from Detmer (1898) *Practical plant physiology*. New York: Macmillan; 8.1b from Bunning (1953) *Entwicklungs- und Bewegungsphysiologie der Pflanze*. Tubingen: Springer, with additional data from Detmar (1898); 8.3 from Salisbury & Ross (1978) *Plant physiology*. Belmont, California: Wadsworth, from data of Zimmer (1962) *Planta* **58**, 288; 8.4a from Salisbury & Ross (1978) *Plant physiology*. Belmont, California: Wadsworth, from data of various authors; 8.4b from Wilkins (1973) *J. Exp. Bot.* **24**, 488; 8.5 from Mansefield & Heath (1963) *J. Exp. Bot.* **14**, 334; 8.7 from Vince-Prue (1975) *Photoperiodism in plants*. Maidenhead: McGraw-Hill; 8.8 from Vince-Prue (1975) *Photoperiodism in plants*.

Maidenhead: McGraw-Hill; 8.9 from Wareing & Phillips (1981) *Growth and differentiation in plants.* Oxford: Pergamon, from data of Hamner (1963). Copyright (1981) by Pergamon; 8.10 adapted from Cumming *et al.* (1965) *Canad. J. Bot.* **43**, 825; 8.11 from King *et al.* (1982) *Plant, Cell & Environ.* **5**, 395; 8.13 from Vince-Prue (1983) in *Encycl. plant physiol.* NS 16B, W. Shropshire & H. Mohr (eds), 457–90. Berlin: Springer, based on data of King & Cumming (1972) *Planta* **108**, 281.

Contents

CONTENTS

CONTENTS

List of Special Topic Boxes

List of tables

Plate 1 Carving from the tomb of Amenophis IV. Akhenaten and Nefertiti make offerings to the solar globe, whose rays end in hands and the hieroglyph for Life; light, the dispenser of life, is transmitted to the royal couple – note also the tropic behaviour of the plant material at the bottom right.

CHAPTER ONE

General introduction

1.1 LIGHT AND LIFE

From primitive times, light has occupied a central rôle in both the physical and spiritual affairs of man (Plate 1). Past cultures seem to have had an acute awareness of the importance of the Sun to life processes – the once-sacred names of some of the ancient Sun gods still persist in modern words (e.g. *Ammon*-ia and, perhaps more fancifully, *Ra*-diation). In fact, it has been claimed that the state of cultural advancement of a society can be equated with its sensibility towards the Sun. In our own society, despite the development of artificial lighting systems, we (and the plant world which supports us) are no less dependent upon sunlight. Even the fossil fuels which are utilized so intensively at present, partly to generate artificial light, represent past entrapments of the Sun's energy.

Light interacts with biological processes in a variety of ways, which can be arbitrarily classified into three general categories. It can have potentially lethal effects on a cell or organism. Some of these can be used to man's advantage – its bactericidal actions and its rôle in the treatment of skin diseases like psoriasis, for example; others are distinctly disadvantageous, such as sunburn, skin-aging, various allergenic effects and even carcinogenesis. Secondly, light is a medium through which an organism can receive information about its environment. Examples of this rôle range from the self-evident such as animal vision, through the less obvious, such as the direction-finding feats of birds and insects, to the very subtle, such as the variety of ways by which light can direct the orientation in space of both motile and non-motile plants. Furthermore, light is the environmental factor uniquely involved in programming the rhythmic processes which regulate the orientation in time of both plants and animals. Thirdly, plants utilize light directly in *biomass production*; through photosynthesis, light constitutes the primary source of all forms of biological energy. These different rôles of light in plant growth and development

1

Table 1.1 Interactions between plants and light.

Type of activity	Phenomenon or response	Organism or example
production of biomass	photosynthesis	bacteria and green plants
orientation in space	phototaxis and photokinesis	motile algae and bacteria
	phototropism	non-motile plants fungi
	heliotropism	sun-tracking by leaves and flowers
	shade effects	growth of stems and leaves
	polarotropism	growth of moss filaments
orientation in time	rhythm-timing	many metabolic activities, cell division and growth, stomatal opening, nyctinastic movements
	photoperiodism	flowering, dormancy induction, leaf abscission, production of tubers, corms, bulbs and runners
determination of form	greening	pigment synthesis and chloroplast development
	effects on growth	hypocotyl hook opening, stem growth, leaf expansion, branching pattern and root growth

are summarized in Table 1.1. (Other possible rôles of light in the origin and selection of early life forms are discussed in Box 1.1.)

The Sun is by far the most important source of biologically significant natural radiant energy. (Suggested effects of moonlight on plant growth have not been substantiated under natural conditions, but there is evidence that starlight is utilized by migratory birds.) The Sun emits a *continuous spectrum* of radiant energies (Fig. 1.1). Approximately half of these are prevented from reaching the surface of the Earth by atmospheric effects, and only the radiant energies with wavelengths between 300 – 1000 nm, the so-called 'biological window', influence life processes. (Nevertheless, the radiation that does pass through the atmosphere is a considerable amount of energy; it is estimated that 28×10^{23} joules of heat equivalent impinge on the surface of the Earth each year.) The radiation that is perceived by our eyes, the visible

2

Box 1.1   The rôle of light in the origin and selection of early life forms

Besides its present influence on biological processes, light must also have played dominant rôles in the origin and selection of early life forms. Radiation is likely to have been a major *energy source* for the synthesis of primitive life molecules. In an atmosphere lacking oxygen, and therefore ozone, much higher amounts of ultraviolet radiation than at present would necessarily have impinged on these molecules. Such dosages would have been likely to result in high rates of chemical change in the early biomolecules, that is, radiant energy is likely to have been a major *mutagenic and selective force*. With the introduction of oxygen into the atmosphere by photosynthesis, and the consequent production of ozone, the amounts of ultraviolet radiation reaching the surface of the Earth itself would have been markedly reduced. Relatively stable replication of the genetic material would have become possible and other forces, as well as light, would assume selective roles.

region, is characterized by wavelengths of 400 – 700 nm, and comprises less than half the energies within the whole biological window. While the energies within this visible region are utilized for the immensely significant process of photosynthesis, plants also make use of radiant energies on either side of the visible range. Therefore it is important to

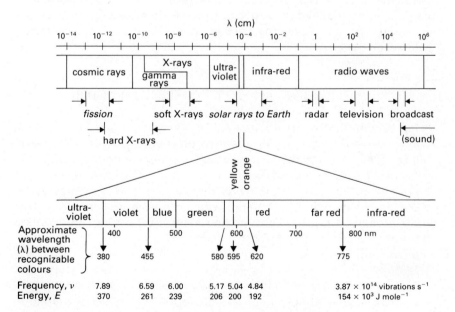

Figure 1.1   The visible portion of the electromagnetic spectrum.

realize that the discrepancy between visible light and the remainder of the biological window can be very misleading when considering the effects of radiant energy on biological materials.

For radiant energy to have any biological effect, whether in the production of biomass or the provision of information, it must first be absorbed by a **photoreceptor**. Although thousands of organic molecules absorb light, only a very few act as biological photoreceptors. The basis for this evolutionary selection resides in the coincidence of:

1 the optical and photochemical properties of an organic molecule;
2 the energy level of the radiation available;
3 the energy required to elicit the biological response.

The energy of radiation is inversely proportional to its wavelength. Radiation of wavelengths greater than 1000 nm has too little energy to cause photochemical change in any available molecules; therefore, no biological work can be obtained from this region of the spectrum. Conversely, radiation of wavelengths less than 300 nm is energetic enough to break chemical bonds, and is therefore highly destructive to putative biological photoreceptors. Thus, molecules have evolved which are capable of absorbing specific levels of radiant energy within the range characterized by wavelengths of 300–1000 nm, and whose subsequent photochemical changes are appropriate to initiate particular biochemical processes. Note that this means that a photobiological phenomenon, initiated by a specific photoreceptor, will be elicited by a specific range of wavelengths: each phenomenon will have a characteristic **action spectrum**.

1.2 LIGHT AS AN ENVIRONMENTAL FACTOR

It has been indicated already that radiant energy is utilized by plants in two quite distinct ways – as a source of energy and as a source of information. Among the many factors of the environment, light is particularly suited for this latter rôle, since it has what may be termed a high information content: unlike factors such as gravity, temperature, water, nutrients, wind, and so on, which can each vary in only two or three dimensions, the light environment can convey information through variations in at least four dimensions: quality, quantity, direction and periodicity.

1 *Quality*. Type of radiant energy, colour, spectral complement, wavelength composition, are all terms which can be used to describe this dimension of the light environment. It can show

4

marked variation between different habitats, for example, in different shade conditions or at different water depths.

2 *Quantity.* Amount of radiant energy, intensity, photon number, fluence rate, specify this second dimension. There can be large differences in the amounts of radiant energy incident on different habitats, at different times of the year, and under different climatic conditions.

3 *Direction.* There can obviously be great variation between different habitats in the direction of irradiation, which gives a spatial dimension to the light environment.

4 *Periodicity.* This describes the regular variation in a temporal dimension. There are, of course, two aspects to this variation: the short-term *daily* cycle of day and night, and the long-term *seasonal* cycle of changing daylength. The characteristics of this periodicity result in the light environment being an important source of temporal information for organisms. Temperature shows regular daily and seasonal fluctuations, but not with the same degree of precision – climatic conditions can result in unseasonally high or low temperatures (there is more 'noise' in the system). Gravitational forces, too, as evidenced by the tides, show regular and predictable fluctuations; but these fluctuations are not in precise phase from year to year (new tide tables are produced for each year). Thus, radiant energy, in terms of the seasonal progression of daylength, is the only reliable environmental or climatic constant. Plants use this 'light clock' to prepare for changes in other environmental conditions. For example, shortening daylength presages the unfavourable growth conditions of winter, and in many species, is the inductive signal to initiate reproductive or regenerative phases of development.

1.3 LIGHT AND PLANTS

The light environment, therefore, besides being a potential source of energy, also contains signals capable of conveying spatial and temporal information. Plants have **photosystems** to process both the energy and the information. There are three parts to a photosystem, each of which must be described before the effect of radiant energy on an organism can be fully understood:

1 *Reception* – the absorption of specific wavebands of energy by a specific photoreceptor which becomes (photo)chemically changed.

2 *Transduction* – the primary action of the photoreceptor, whereby the radiant energy is transformed into a relatively stable

biochemical form of energy; at this stage in the processing of information signals, there is also *amplification* of the transformed signal.

3 *Response* – the modulation of metabolism to result in an event which is observable at the physiological level; note that this modulation must occur within genetic constraints, and will be appropriate to the overall adaptive behaviour shown by a particular organism. That is, *the same environmental signal can provoke different responses in different organisms* (e.g. sun and shade plants show different responses to shade conditions).

In plants, there are at least three types of photosystem. As yet we understand their respective modes of operation (in terms of receptor action, transduction event, and metabolic response) to markedly different extents.

Photosynthetic systems harvest radiant energy and transform it into chemical energy. The receptor pigments for the process are various chlorophylls and carotenoids. This combined operation of a variety of pigment types, each with its characteristic absorption properties, ensures that energy from a wide proportion of the spectrum can be harvested. The receptors are well characterized as pigments, the primary transduction act is known, and the subsequent metabolic events which lead to the production of a stable form of chemical energy are reasonably well understood. Since this activity is concerned (simply) with energy harvest over a wide region of the spectrum, it is the *quantity* of light which is the significant parameter.

The **phytochrome photosystems** are involved in monitoring and processing information carried by the light environment. The photoreceptor is fairly well characterized as a pigment, and a very large number of physiological responses are known to be initiated by its action. Despite this, neither the primary transduction step nor the subsequent metabolic events are at all clear. Phytochrome exists in two isomeric forms which are interconvertible by radiant energy in the red (R) and far-red (Fr) regions of the spectrum. The far-red absorbing isomer (P_{fr}) is thought to be the biologically active form. However, phytochrome does not function as a simple on–off switch; in nature, it is exposed simultaneously to the different wavelengths which constitute daylight. The action of phytochrome is related to the particular equilibrium between the two isomeric forms, which in turn derives from the ratio of red : far-red radiation in the spectrum. Therefore, a fundamental function of phytochrome is to monitor the *quality* (R : Fr) of the light environment. It may also be involved in detecting *quantity* of light. Furthermore, since phytochrome interacts with many endogenous rhythms in plants, its action also provides a means for timing the *periodicity* of irradiation.

The third type of photosystem is represented by the specific responses to blue light that are shown by many organisms, including plants. In most cases, the chemical nature of the **blue-light photo-receptors** is not established, and in fact, many authorities consider it likely that more than one type of blue-light receptor exists. Certainly, the responses to blue light vary greatly in type, and include effects on endogenous rhythms, organ orientation, stem extension, stomatal opening and cytoplasmic streaming. Many of the responses to blue light also show a strong dependence on the amount of light. Therefore these blue-light receptors seem to be variously involved in monitoring the *quality*, *quantity*, *direction* and *periodicity* of irradiation. (The term **cryptochrome** is sometimes used to denote photosystems that are triggered by blue light; the name originates from a pun on Cryptogam, the order of lower plants in which a non-phytochrome photoreceptor was first suspected. However, use of this single term probably implies too strongly that there is a single type of blue-light receptor.)

Thus, within the biological window of the Sun's spectrum, plants can obtain energy to some extent from the entire visible spectrum, but obtain information particularly from the blue, red and far-red regions. In photosynthesis, the metabolic events are initiated by the photo-oxidation of a special pigment type at the reaction centre of the photosynthetic unit. The light-absorbing pigments are present in high concentrations (approximately 25 mM per chloroplast), and the rate of the process is directly proportional to the photon fluence rate, up to relatively high light intensities. In contrast, the processes of photo-

Box 1.2 Radiant energy and the determination of plant form

Besides its direct effects in photosynthesis and photomorphogenesis, radiant energy has played another significant, though indirect, rôle in the determination of plant form. Plants are, of necessity, exposed to large amounts of radiant energy. However, in the majority of plants, only a small fraction of the radiant energy that is absorbed can be utilized in photochemical work. The rest of the energy must be dissipated as heat and returned to the environment. Many features of plant design (e.g. shape and orientation of leaves) aid this heat loss. However, a design which enhances heat loss during the day may also be a disadvantage at night. Indeed, in some situations, there is a real danger of supercooling, with 'frost damage' to tissues occurring at environmental temperatures above the freezing point. The folding at night of the leaves of certain species ('sleep movements') – interestingly, itself a light-triggered phenomenon – was originally considered by Darwin to be an adaptation for heat conservation.

Table 1.2 General sensitivities of photobiological processes to different intensities of natural light (from Bjorn 1976).

Light intensity (W m^{-2})	Equivalent	Process
1000	sunlight, noon, summer	photosynthesis (saturates at 200–300 W m^{-2})
100	daylight, cloudy	
10		photosynthesis (compensation point)
1		flowering
0.1	twilight	seed germination
0.01	moonlight	
0.001		phototaxis (algae) colour vision (man)
0.0001		plant greening
0.00001		phototropism (coleoptile)
0.000001		phototropism (many fungi)
0.0000001		black and white vision (man)
0.00000001		
0.000000001	starlight	
0.0000000001		growth inhibition (dark-grown oats)

$$10^{-12} \text{ W m}^{-2} = \text{limit of human eye}$$
$$10^{-19} \text{ W m}^{-2} = \text{limit of largest telescope}$$

morphogenesis, which involve mechanisms of signal amplification, can be initiated by very low photon fluence rates (Table 1.2). Furthermore, since photomorphogenic responses are concerned with the orientation of a plant in space and time, they frequently show extremely complex interactions with the amount, duration and timing of irradiation. Other indirect relationship between plants and light are discussed in Box 1.2.

1.4 LIGHT AND PHOTOMORPHOGENESIS

It could be argued that the terrestrial green plant interacts with its physical environment at a more intimate level than do most animals. Not only is the plant totally dependent on this environment for its supply of energy, but also, being non-motile, it can only cope with the rigours of the environment by appropriate adjustment of its growth and development. Since the plant is so completely dependent on radiant energy, it is hardly surprising that this factor should exert major effects on plant form.

During its initial growth, whether from a seed or other organ of perennation, the plant exists heterotrophically on stored food reserves.

Upon exposure to a significant amount of light, it becomes transformed into an independent autotrophic organism (Fig. 1.2). The most obvious events during this process of **de-etiolation** are related to the rapid promotion of a photosynthetic capacity. These include changes at the morphological level (e.g. deceleration of stem extension and promotion of leaf expansion); at the cellular level (e.g. chloroplast development); and at the molecular level (e.g. pigment and enzyme synthesis). Light itself is the initiating force in all these processes. Phytochrome plays a major rôle in many of them, but other photoreceptors are also involved. For example, many of the effects of light on chloroplast differentiation and stem extension seem due to the action of blue-light receptors. Protochlorophyllide, a precursor of chlorophyll, is a photoreceptor for the light induction of further chlorophyll synthesis. Thus, the process of greening is a complex series of events which involves the integrated operation of many photosystems.

Less obvious changes which also occur during de-etiolation include enhancement of the plant's ability to respond to informational signals in the light environment. During the transformation into an autotrophic organism, not only is a photosynthetic ability developed, but also the photomorphogenic capacity itself is enlarged. In etiolated tissues, major systems responsible for the orientation of a plant in space and time are not fully developed; the etiolated seedlings of many dicotyledonous species do not have a strong phototropic response; etiolated plants generally do not show any response to daylength. An important aspect of the enhanced photomorphogenic responsiveness that occurs as the plant becomes an autotrophic organism is the entrainment, again by light itself, of the endogenous rhythms whose further interactions with specific light regimes form the basis of time-measuring ability.

1.5   LIGHT AND ORIENTATION IN SPACE

In some primitive micro-organisms, a limited degree of spatial orientation is achieved through the photosynthetic process itself. For example, in the purple bacterium *Chromatium*, a sudden decrease in light intensity brings about a transient drop in the supply of biochemical energy to the flagellar apparatus. Thus, if the organism reaches the edge of an irradiated area, active bacterial motion comes to a temporary halt. When flagellar activity is resumed, the cell may be pointing in a different direction. This simple stop–go mechanism gives the organism a high probability of remaining in an irradiated area, although it does not involve a positive response to the direction of irradiation. The navigation system is much more refined in those green

(a)　　　　　　　　　　　　　　　(b)

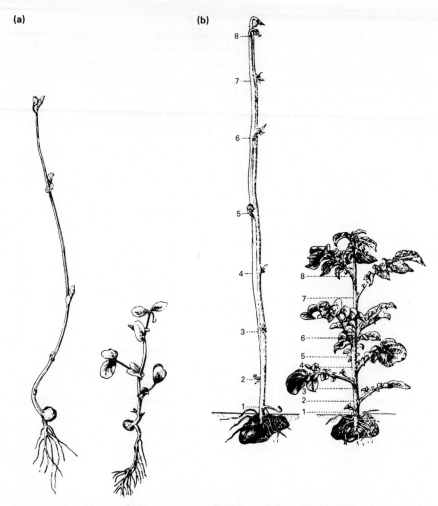

Figure 1.2 Effects of light on plant morphology. (a) Pea seedlings grown in darkness (left) or in light, for six days. (b) Potatoes sprouted in darkness (left) or in light.

flagellates, such as *Euglena*, where there is a mechanism for actually determining the direction of irradiation. In such organisms, at the base of the flagellum there is a carotenoid pigment complex which controls flagellar activity. The complex consists of a photoreceptor and a screening pigment, the *eyespot*. The screening pigment casts a shadow on the photoreceptor and thus, flagellar activity is regulated according to the relationship between the cell and the direction of irradiation. Such movement of an organism, truly in relation to the direction of light, is termed **phototaxis**: if the rate of movement is also affected by light, this phenomenon is called **photokinesis**. Phototaxis can be

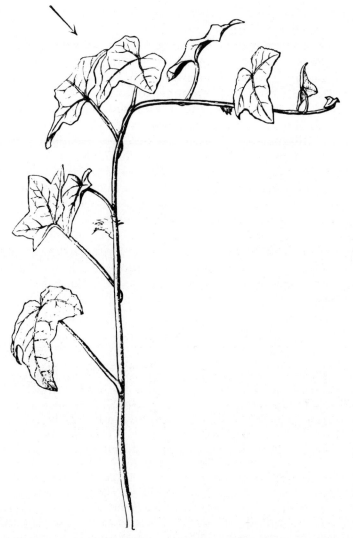

Figure 1.3 Phototropic response in ivy (*Hedera helix*). Note that while the main stem bends away from the light, the leaf petioles curve towards it.

positive or negative (i.e. towards or away from the source of irradiation), even within one species, depending on such factors as light intensity and nutritional conditions.

Non-motile plants, of course, respond to environmental stimuli by regulation and adjustment of their growth. **Phototropism**, which can also be positive or negative (Fig. 1.3), is a directional growth movement in response to the direction of irradiation. The action spectra for such responses are similar over a wide range of organisms, from fungi to higher plants, and indicate that phototropic responses are generally mediated through the actions of blue-light receptors. Phototropism is most apparent in young, rapidly growing tissues; a difference in the rates of growth on the two sides of an organ results in curvature of the organ towards, or away from, the light source. Most aerial shoots show a positive phototropic response, while only some roots are negatively phototropic. Directional responses of plants to the plane of polarization of light are discussed in Box 1.3.

The term **heliotropism** is used to describe the Sun-tracking movements of certain plant organs, such as the leaves of *Tropaeolum majus* (nasturtium) and the flowers of *Helianthus annus* (sunflower). (In one sense, the phenomenon is technically not a tropism, since in most cases the actual movement arises from turgor changes in the motor cells of a pulvinus rather than from differential growth; however, there is response to the direction of stimulus, which also distinguishes the phenomenon from a nastic movement.) In the few cases investigated, the most effective part of the spectrum to produce the Sun-tracking response is in the blue region. The function of Sun tracking is different in different species. Where leaves are maintained in an orientation at

Box 1.3   Polarotropism

Plants can show directional responses to the plane of polarization of light – a phenomenon known as polarotropism. Filaments from germinated fern spores grow at right angles to the plane of vibration of the electrical vector in linearly polarized light; if the plane of vibration is changed, the direction of growth changes. In *Dryopteris*, the action spectrum suggests the involvement of both phytochrome and a photoreceptor for blue light; in *Sphaerocarpus*, the response is activated only by blue light. The phenomenon is due to the *dichroic* orientation of the photoreceptors on a cell membrane (dichroism = differential absorption of light which is plane polarized in different directions). These findings have been derived from laboratory investigations, where linearly polarized light has been used in attempts to determine the intracellular location of phytochrome. At the moment, there is no clear indication that the phenomenon is important in the field. It is difficult to demonstrate polarotropism in multicellular organisms: there is no physiological response which allows detection of dichroic orientation of the photoreceptor. However, it may be of interest to note that a significant proportion of reflected light in the environment is polarized. Further, when maize coleoptiles are exposed to linearly polarized red light, more conversion to P_{fr} occurs when the plane of vibration of the light is perpendicular to, rather than parallel with, the long axis of the organ. (See Fig. 5.3 and Fig. 7.3.)

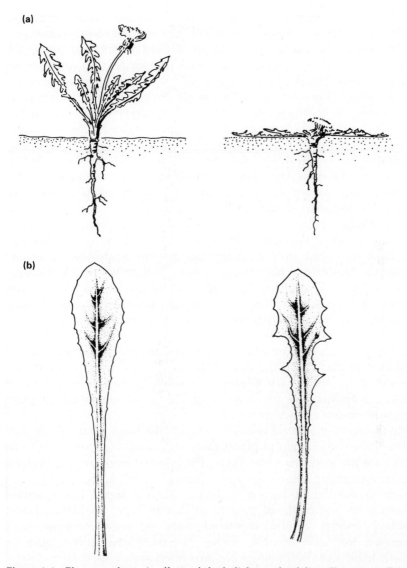

Figure 1.4 Photomorphogenic effects of shade-light on dandelion (*Taraxacum officinale*). (a) Rosettes grown in shade (left) or in open conditions. (b) Leaves from plants grown wholly in 10-hr periods of white light (right) or subjected to a few minutes far-red irradiation at the end of the normal light period.

right angles to the direction of irradiation (*diaheliotropism*), light interception is maximized; rates of photosynthesis are markedly enhanced, particularly in the morning and evening when the Sun is at low angles of elevation. In other species, the leaves are aligned parallel to the Sun's rays (*paraheliotropism*), and the heat load is thus minimized. A phenomenon of similar adaptive value, but very different mechanism, is seen in the behaviour of the so-called compass plants; here, the leaves are held in a fixed orientation along the meridian. For example, in *Lactuca scariola* (prickly lettuce), the laminae face east–west. The decreased light interception at midday reduces leaf surface temperature by up to 5°C, while photosynthetic rates remain high during the morning and afternoon. Such irreversible orientation is fixed during leaf development, but the mechanism seems to involve photoperception, since plants grown in the shade have randomly orientated leaves. Another function of solar tracking is illustrated by the flowers of certain Arctic species: inward reflection of radiant energy by the petals measurably elevates the temperature within the flower, thus aiding pollination and seed development.

Phytochrome is not directly involved in phototropism, although its action can change the sensitivity of an organ to blue light. However, red light itself can induce directional growth responses in certain situations. Protonemal filaments of a moss will grow towards a source of red light, as a result of light-induced changes at the growing tip. Further, if a seedling of a higher plant (e.g. cucumber) is grown under red light, with one of its two cotyledonary leaves covered, the seedling will eventually curve in the direction of the remaining irradiated leaf (see Fig. 7.7); this response has been related to a phytochrome-induced inhibition of growth on the side of the stem below the irradiated leaf. The extent to which such growth responses, in contrast to blue-light induced tropisms, are involved in the general orientation of plant organs in nature is undetermined.

Phytochrome-mediated responses which do have great significance to the spatial orientation of plants (although not in a strong directional sense) are the so-called *shade effects*. The spectral environment under a vegetation canopy differs markedly from daylight, in having a much higher proportion of far-red radiation. Such a signal has a strong influence on the mode of growth of certain types of plant: shade-avoiding species exhibit increased rates of stem extension and decreased leaf development. Other effects of shade-type light on growth habit and morphology are also often seen (Fig. 1.4).

1.6   LIGHT AND ORIENTATION IN TIME

There are two major aspects to the temporal behaviour of plants: first, there are the responses of a plant to the *daily* cycle of light and darkness, during which it must alternate between autotrophic and heterotrophic modes of growth; secondly, there is the position of a plant in the *yearly* cycle of changing climatic conditions. Light signals are deeply involved in both these aspects of temporal behaviour. However, in all cases, the environmental signals interact with some form of endogenous rhythm. The interaction may consist of the triggering or entrainment of rhythmic behaviour by, say, the signals contained in the particular light environments of dawn and dusk; or it may take the form of coincidence of a specific light signal (e.g. a particular daylength) with an appropriately sensitive phase of growth.

The behaviour of stomata provides an example of a daily cycle of response. The stomatal mechanism is based upon turgor changes of the guard cells, probably as a consequence of changes in their concentration of potassium ion. Many factors influence these changes in turgor, including an endogenous rhythm to guard-cell swelling and shrinkage, an effect of CO_2 on shrinkage (pore closure) and effects of light on swelling (pore opening). The dominance of each of these factors varies according to the species. The light effects are again via the red and blue regions of the spectrum, although in this case it is thought that red light acts through the photosynthetic pigments, while blue light acts through a specific blue-light receptor. Therefore, in the natural environment, stomatal responses are regulated by an endogenous rhythm whose actions are reinforced by light.

Another daily cycle of behaviour is illustrated by the **nyctinastic movements** of certain leaves (Fig. 1.5). During the day, the leaves are oriented more or less horizontally, but at night they fold in against the main body of the plant (upwards or downwards, depending on the species). Again the basis of the movement is a turgor change induced by potassium flux, in this case in special motor cells at the base of the leaf. The endogenous rhythmic component is more dominant in this cycle, but light signals are still involved in both entraining and reinforcing the response.

An even more complex interaction of the light environment with the temporal behaviour of plants is seen in the phenomenon of **photoperiodism**. This is defined as a non-directional developmental response to the non-directional but periodic stimulus of daylength. It is the major means by which the various developmental phases of the plant are related to the seasonal growing conditions of the environment, and is of particular significance to those phases of growth concerned with floral and vegetative modes of reproduction and regeneration. Photoperiodic behaviour is thus seen in many aspects of plant development – for example, in flowering, bud development, stolon and runner production, and in the formation of perennating organs such as corms, bulbs and tubers. It is especially apparent in plants which grow in environments where there are distinct seasonal climates. In temperate regions, the production of all forms of perennating storage organs

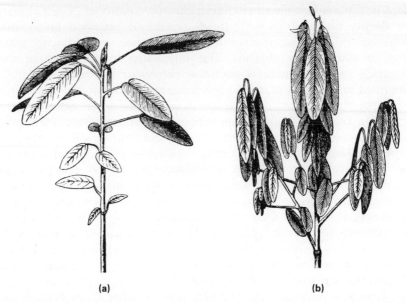

(a)   **(b)**

Figure 1.5 Nyctinasty (sleep movements) in *Desmodium gyrans*. Leaf orientation during the day (left) and at night.

(except, curiously, onion bulbs) is induced by the shortening daylengths of an approaching winter. However, photoperiodism in plants has been most intensively studied in relation to flowering. Again, the actual type of response is related to the overall adaptive behaviour of a particular species. Short days, presaging adverse climatic conditions, induce flowering in some species; but in the floor species of a deciduous forest, there may be greater adaptive value in flowering being induced by the long days of approaching summer, with the result that reproduction is completed before the forest canopy closes over. The mechanisms of photoperiodic control are not known. Phytochrome obviously plays a rôle as a photoreceptor, although it may not be the only one involved. Even the nature of the signal which specifies the length of day is somewhat controversial, between quantity or quality of radiation. Thus, the phenomenon of photoperiodism, despite being of major importance, is poorly understood at almost all points in our concept of a photosystem: signal, receptor, primary event, metabolic change and physiological response.

1.7 SUMMARY

1 Plants use light as an energy supply and as a source of information.

2 The light environment can vary in quality, quantity, direction and periodicity of irradiation. Periodicity, in terms of daylength, is the major environmental clock and calendar.

3 Any effect of light operates through a specific photoreceptor. Each photoprocess thus has a specific action spectrum.

4 Plants have three types of photosystems for processing the energy and informational signals of light: the photosynthetic pigments, phytochrome and blue-light receptors. Their respective modes of operation, in terms of reception, transduction (amplification) and response, are understood to different extents.

5 Energy can be harvested throughout the visible region of the spectrum: information is provided almost exclusively by the blue, red and far-red regions. Photomorphogenic effects are initiated by much lower photon fluence rates than is photosynthesis.

6 Radiant energy exerts major effects on plant form and behaviour; these include development of photosynthetic ability, entrainment of endogenous rhythms and orientation in space and time.

7 Mechanisms which orient plants in space include phototaxis, photokinesis, phototropism, heliotropism and shade effects.

8 Temporal responses operate through the interaction of light signals with biological rhythms; they include daily responses and photoperiodic responses to seasonal change in daylength.

FURTHER READING

Bjorn, L. O. 1976. *Light and life*. Sevenoaks: Hodder & Stoughton.

Kendrick, R. E. and B. Frankland 1976. *Phytochrome and plant growth*. London: Edward Arnold.

Mohr, H. 1972. *Lectures on photomorphogenesis*. Berlin: Springer.

Smith, H. 1975. *Phytochrome and photomorphogenesis*. Maidenhead: McGraw-Hill.

Vince-Prue, D. 1975. *Photoperiodism in plants*. Maidenhead: McGraw-Hill.

Whatley, J. M. & F. R. Whatley 1980. *Light and plant life*. London: Edward Arnold.

CHAPTER TWO

Radiant Energy

2.1 NATURE OF RADIANT ENERGY[1]

Radiant energy exists in the form of the electromagnetic spectrum, which extends from cosmic rays to radio-waves; the visible region is only a narrow band within the whole energy spectrum (Fig. 1.1). Several phenomena, including those of diffraction and interference, indicate that radiation is propagated as waves. In this form, it is depicted as oscillating, transverse electric and magnetic vectors (Fig. 2.1) which are characterized by wavelength, λ (lambda), the length of a cycle; and frequence, ν (nu), the number of cycles per second. The velocity, c, is constant (speed of light in a vacuum = 2.998×10^8 m s^{-1}), and is the product of the wavelength times the frequency:

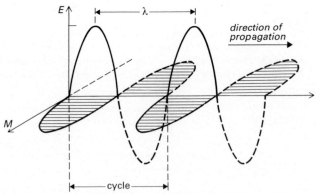

Figure 2.1 Diagrammatic representation of a light wave, showing the simultaneous electric (E) and magnetic (M) vectors as a function of position along the axis of propagation.

[1] See Bickford & Dunn (1972), Clayton (1970), Salisbury & Ross (1978), Seliger & McElroy (1965).

$$c = \lambda \times v \ ; \qquad v = \frac{c}{\lambda} \ ; \qquad \lambda = \frac{c}{v}$$

However, radiation interacts with matter as whole units of energy, and thus also shows the properties of particles. Such a single pulse of electromagnetic energy is called a quantum (or **photon**, if it is in the visible region of the spectrum). The energy of a quantum is a function of its frequency and is therefore inversely proportional to its wavelength:

$$E = hv = \frac{hc}{\lambda}$$

where E = energy of a quantum or photon, and h = Planck's constant (= a universal constant which relates the energy of a photon to the frequency of the radiant energy; its dimensions are energy and time and it is equal to 6.6×10^{-34} J s.) The radiant energies of different wavelengths are shown in Table 2.1.

It should be particularly noted that:

1 light is transmitted as waves, but it *interacts with matter as particles;*
2 light is *described by its wavelength,* usually given in nanometres;
3 *the energy of a photon is inversely proportional to its wavelength.*

2.2 SOURCES OF RADIANT ENERGY[2]

Radiant energy is emitted as a consequence of various processes. *Thermal radiation* is emitted by virtue of the temperature of a substance, whereby the kinetic energy of heat motion of the molecules is dissipated by emission of a continuum of radiation. Most solids emit visible radiation (i.e. wavelengths less than 780 nm) at temperatures above 500°C. A characteristic of radiation from such a source is that it exists as a *continuous spectrum* of energies, the distribution of which is determined by the temperature of the emitting substance (Fig. 2.2); the higher the temperature, the more the spectrum is shifted downwards into the visible region (substances first become red-hot, then white-hot).

Radiation can also be emitted as a result of an *electric discharge* through a gas. Individual molecules in the gas are electrically excited to a higher energy level, and the 'excitations' are released as quanta of

[2] See Bickford & Dunn (1972), Clayton (1970), Seliger & McElroy (1965).

Table 2.1 Radiant energies of different wavelengths.

Wavelength (nm)	Colour	Energy level (joules per μmole)[a]
300		0.399
400	violet	0.299
500	green	0.239
600	orange	0.199
700	red	0.171

Note: Energy of a quantum $= \dfrac{2 \times 10^{-16}}{\lambda \text{ nm}}$ joules.

[a] A mole of light $= 6.023 \times 10^{23}$ photons.

radiant energy. In this case, the radiation is emitted from a specific energy level and will thus be of a specific wavelength. A discharge source of radiant energy is characterized by a *line spectrum* which is superimposed on the background temperature continuum.

Chemiluminescence is the emission of radiant energy from an energy-releasing chemical reaction; it, too, is of specific wavelengths, characteristic of the particular change in energy levels in the reaction. Photoluminescence is the re-emission of radiant energy from molecules which themselves have been photostimulated; *fluorescence* is one form of this phenomenon. Finally, radiant energy can be emitted in response to local electric fields, which can be generated in a variety of

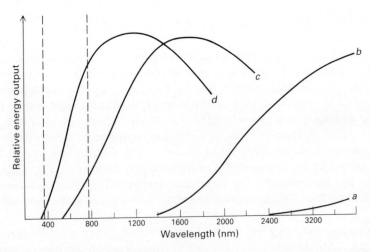

Figure 2.2 Blackbody radiation curves for different operating temperatures, (the visible region is indicated within the dashed lines). *a* – steam radiator (100°C); *b* – heater (500°C); *c* – infra-red source (melting iron, 1500°C); *d* – incandescent lamp (filament, 2700°C).

ways (e.g. triboluminescence describes the sparks produced by mechanical breakage of crystalline materials).

2.3   INTERACTION OF RADIANT ENERGY. WITH MATTER[3]

In terms of quantum mechanics, photons and electrons can both be described by wave functions. An atom is considered to have its electrons in different orbitals, which are defined by their angular velocities and their energies. Different orbitals have different possible energy levels or permissible states; the lowest energy level is referred to as the ground state, the others as **excited states**. Between these permissible states, there will be specific and characteristic differences in energy level (ΔE). For absorption of radiant energy to be possible, the difference in energy levels between the ground state of a molecule and an excited state must match the energy content of a photon:

$$\Delta E = h\nu = \frac{hc}{\lambda}$$

where E = energy level of an orbital. This means that any one type of molecule can only absorb certain wavelengths of radiant energy; it has a characteristic **absorption spectrum**. Similarly, emission of radiant energy represents a specific drop in energy level within a molecule, and consists of wavelengths characteristic of the specific state changes of that molecule. The situation is illustrated in the diagrams of Fig. 2.3. In the well spaced atoms of a gas, the permissible states have well defined, distinct energy levels. In such media, absorption and emission are characterized by *line spectra*, that is, they involve discrete levels of radiant energy. In molecules and solutions, the energy levels representing excited states are broadened by the existence of substates, due to atomic and molecular interactions. Therefore, the ranges of energy levels which can be involved in state changes cover *spectral bands* of wavelengths. The ground state of a chlorophyll molecule, for example, can be raised to one of two possible excited states; the changes in energy level involved in attaining these states correspond, respectively to photon energies in the blue and red regions of the spectrum. The excited state reached by absorption of the more energetic photons of the blue region is at a higher energy level than that brought about by red irradiation; but photons in the green region of the spectrum, although intermediate in energy level between blue

---

[3] See Clayton (1970), Hayward (1984), Schäfer & Fukshansky (1984), Seliger & McElroy (1965).

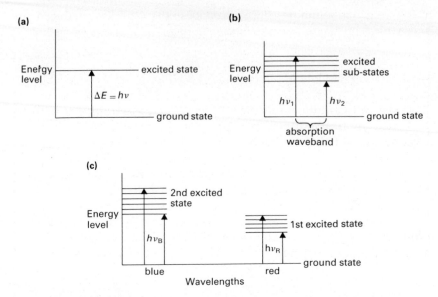

Figure 2.3 Diagrammatic representations of changes in energy levels of electron orbitals after photon absorption: (a) atoms in a gas; (b) molecules in solution; (c) excitation of chlorophyll.

and red, do not have energies specific to the change required to reach either excited state, so cannot be absorbed.

Thus, radiant energy is absorbed as whole, discrete quanta, each absorption causing a change in the nuclear–electronic equilibrium of an absorbing molecule. This change in state results in the molecule becoming more chemically reactive. There are two features of radiation which must be clearly distinguished at this point. The **energy** of specific types of photons determines *whether* particular photons can be absorbed by a particular molecular species; it is measured in energy units (joules). However, it is the *number* of photons actually absorbed which determines the *number* of molecules that are photochemically excited. Therefore, it is the **photon fluence rate** (number of photons per unit time per area) which determines the amount of (photo)chemical change. Further discussion of the interaction of radiant energy with matter is given in Box 2.1.

In different types of molecule, the excitation state can be dissipated in different ways (Fig. 2.4). It can be lost by thermal relaxation (light can heat substances): in this case, the nature of the irradiated material is such that its permissible states have overlapping energy levels through which the excitation energy can 'gradually' decay down to the ground state (Fig. 2.4a). In another type of molecule, the excitation energy may be re-emitted as *photoluminescence*. Here, there will have

Box 2.1 The laws of photochemistry

The laws of photochemistry give formal expression to the interactions between radiant energy and matter (although the complexity of biological systems usually makes it difficult to apply these laws with any great degree of rigour to the analysis of biological responses). The first three laws pertain to the absorption of radiation; the others are concerned with its effects on matter.

(1) *Grotthus & Draper (1818)*: only light which is *absorbed* can produce a photochemical change. (Application of this law allows comparison between the absorption spectrum of a putative photoreceptor and the action spectrum for a photobiological response.)

(2) *Bunsen & Roscoe (1850)*: duration and intensity of irradiation show *reciprocity* in level of effectiveness – that is, if the arithmetical product of irradiation intensity × duration is held constant, then brief irradiations at high intensity produce the same level of effect as longer irradiations at low intensity. (If such a relationship is found in a response, it indicates that only one type of photoreceptor is involved.)

(3) *Beer & Lambert (1860)*: solutes absorb a *constant fraction* of the radiant energy passing through a solution. (This means that radiation passing through a solution will show an exponential, rather than linear, decline. Further, application of this law to measurements of light transmission through a solution allows determination of pigment or solute concentrations.)

(4) *Planck (1900)*: radiant energy exists as discrete indivisible *quanta*.

(5) *Stark & Einstein (1905)*: The *law of photochemical equivalence* states that the absorption of one quantum brings about a photochemical change in one molecule (or atom or electron). (It is this statement which is crucial to an understanding of initial events in a photobiological response.)

The *quantum yield* of a reaction, Φ (capital phi), relates the extent of response to the number of photons absorbed:

$$\Phi = \frac{\text{number of molecules affected}}{\text{number of photons absorbed}}$$

The Stark–Einstein relationship indicates that in a strictly photochemical reaction, $\Phi = 1.0$. However, quantum yields greater than 1.0 are not uncommon – for example, in a photo-initiated chain reaction, or in a photobiological response that has been subject to metabolic amplification (Ch. 5).

Note that the quantum yield is calculated from the number of photons actually absorbed, a parameter that is usually very difficult to determine. The terms 'relative quantum efficiency' or 'photosensitivity' relate the response to the more readily measurable parameter of the number of photons incident on the material:

$$\text{relative quantum efficiency} = \frac{\text{number of molecules affected}}{\text{number of photons incident}}$$

The term *photoconversion cross section*, σ (sigma), is directly comparable with this, and describes the probability that a photon will bring about the particular photoreaction. The 'cross section' is itself determined by two factors: the absorption characteristics of the absorbing molecule (its absorption coefficient, ε (epsilon), and the quantum yield of the reaction:

$$\sigma = \varepsilon \Phi$$

Both σ and ε are measurable characteristics of a photoreactive molecule (see Hayward 1984).

been some loss of energy by thermal decay down through substates – that is, the re-emitted radiation will have a lower energy level; fluorescent radiation is always of longer wavelengths than the exciting radiation (Fig. 2.4b). In both these cases, the excitation energy is lost from the system – no useful work is obtained from the radiant energy.

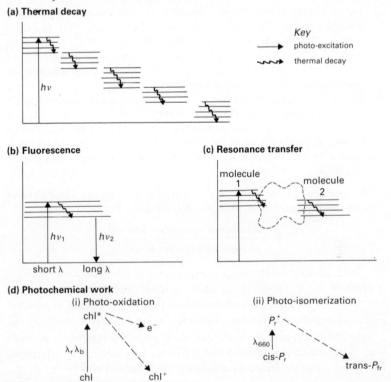

Figure 2.4 Diagrammatic representations of the various ways in which photo-excitation energy can be dissipated.

In certain situations, the excitation energy can be passed from the original photoexcited molecule to another molecule by the process of *resonance transfer* (Fig. 2.4c). For this, at least two criteria must be satisfied: the substate energy levels of the molecules must overlap; and the physical arrangement between the molecules must be highly ordered. Again, because of energy loss by thermal decay, such transfer can only occur from molecules capable of absorbing shorter wavelength (higher energy) radiation to molecules which themselves would be excited by longer wavelength radiation. (This phenomenon is of importance in photosynthesis, where the radiant energies harvested by various pigment types are 'channelled' to chlorophyll reaction centres.) Finally, the excitation energy can be used to carry out *photochemical work*: the excited, more reactive molecule undergoes a chemical reaction. Many types of photochemical reaction are known, but two are of particular biological significance (Fig. 2.4d). In photosynthesis, the primary event is a photo-oxidation: the excited chlorophyll molecule loses an electron, the subsequent effects of which enable other metabolites to be synthesised. In phytochrome-mediated responses, the primary event may be a photo-isomerization: it is thought that the excited phytochrome molecule undergoes a configurational change. This results in a change in its biological activity, and sets in motion the sequence of events responsible for the metabolic amplification of the initial light signal.

2.4 MEASUREMENT OF RADIANT ENERGY[4]

2.4.1 *Quantity*

Light quantity can be measured in three different ways: photometrically, radiometrically and in quantum terms. Each of these represents a different basis of measurement and *each is concerned with a different property of radiant energy*. Not all of these measurements are appropriate for studying the light responses of plants.

A **photometric** basis of measurement is used by lighting engineers and in camera light meters. The sensor of a light meter consists of a photosensitive cell which converts radiant energy into an electric current. However, the photocell, and the series of filters around it, are designed so that the spectral sensitivity of the instrument matches that of the human eye, with maximum sensitivity in the green region of the spectrum at λ 550 nm (Fig. 2.5). Such an instrument, therefore, measures brightness or luminosity, not energy or photons. The basic unit of luminosity is the lumen (flux emitted per unit solid angle by a

[4] See Holmes (1984b), Seliger & McElroy (1965), Withrow (1943).

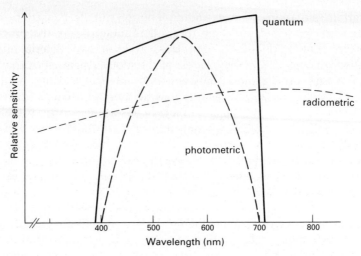

Figure 2.5 Relative sensitivities of different types of photosensor used in the measurement of 'light'.

point source of 1 candela). Terms derived from the lumen to describe 'illumination' (i.e. luminosity per area), are the lux, which equals 1 lumen per square metre, and the foot candle, which equals 1 lumen per square foot. Because these units are really based upon the sensitivity of the average human eye, rather than on that of plant pigments such as chlorophyll, phytochrome or a blue-light receptor, *photometric methods are totally inappropriate for the investigation of the effects of radiant energy on plants.* Indeed, they can be seriously misleading. Equal luminosities at different wavelengths can represent markedly different amounts of energy:

$$1 \text{ lumen at } \lambda \ 562 \text{ nm} = 1.49 \text{ mW}$$
$$1 \text{ lumen at } \lambda \ 430 \text{ nm} = 126.7 \text{ mW}$$

That is, in the less visible region of the spectrum (λ 430 nm) much more radiant energy is required to give the same level of brightness. Therefore, photometric methods cannot be applied with any meaning to comparisons of sources with different spectral distributions (including comparisons of 'daylight', in environments where light quality may be different).

Radiometric methods measure energy in jourles, or, when integrated over time, watts (1 W = 1 J s^{-1}). The measuring instrument, a radiometer or solarimeter, incorporates a thermopile across which radiant energy generates a temperature-induced electromotive force (EMF). Such a device is therefore sensitive to all forms of radiant energy in the daylight spectrum, including infra red radiation (Fig. 2.5). Measurements of visible radiation can be obtained by using appropriate filters. (A KG2 heat filter blocks off infra red radiation; alternatively, Wratten 88 or RG2 types of filter transmit only infra red radiation.) These methods do have use in plant research, particularly in physio-ecological investigations of photosynthesis, where it is often the total PAR (photosynthetically active radiation, λ 400–700 nm) which is of interest; radiometric measurements allow energy input to be related to productivity. However, radiometric methods should be used with caution in photomorphogenic research. They measure total radiant energy and take no account of the different energies of different wavelengths. Therefore, they should not be used to compare heterochromatic sources of radiation whose spectral distributions may be significantly different. More important, radiometric methods alone should not be used in comparisons of the relative physiological effectiveness of different spectral regions on some response: if, say, blue and red light treatments were balanced in energy terms, then, because of the inherently greater energies of 'blue photons', more photons would be present in the red light treatments; yet it is the number of photons which determines the amount of photochemical change. (If only radiometrically based instrumentation is available, a simple calculation can be used to convert energy units into numbers of photons – see below.)

Quantum methods make use of an instrument called a quantum sensor to measure the number of photons. Like a light meter, a quantum sensor consists of a photovoltaic cell, by which radiant energy is transformed into an electric current. However, the sensor also incorporates a set of filters which 'quantum correct' the response between λ 400–700 nm (Fig. 2.5), so that it gives a direct response to number of actual photons (integrated over time and area). The unit of measurement is the **mole**: 1 mole of light is Avogadro's number of photons (6.023×10^{23}). (In older texts, a mole of light may be referred to as an 'einstein'). The choice of the mole as the unit derives directly from the Stark–Einstein law of photochemical equivalence (1 photon : 1 electron): thus, 'amount of light', as number of photons, can be directly related to 'amount of chemical change', as number of molecules. Quantum measurements, therefore, are the most appropriate for investigation of photobiological effects. (The quantum sensor was originally developed for photosynthetic studies. These models, 'PAR-sensors', have relatively sharp cut-offs in sensitivity at λ 400 nm and 700 nm, and cannot measure far-red radiation. Instruments for measuring photomorphogenic radiation, R-sensors and Fr-sensors, are available.)

Table 2.2 Terminologies and units used in radiant energy and quantum measurements.

Parameter	Term	Energy units (joules)	Quantum units (μmoles)	Example of use
quantity per area	fluence	$J\ m^{-2}$	$\mu mol\ m^{-2}$	total amount received by a leaf.
quantity per time	rate	$J\ s^{-1}$(watt)	$\mu mol\ s^{-1}$	flow rate, output from a lamp.
quantity per area per time	fluence rate	$W\ m^{-2}$	$\mu mol\ m^{-2}\ s^{-1}$	flow rate per area, irradiation of a surface.
quantity per spectral distribution	spectral fluence rate	$W\ \lambda^{-1}\ m^{-2}$	$\mu mol\ \lambda^{-1} m^{-2} s^{-1}$	description of a spectrum.

'Amount of light', in energy or quantum measurements, can be expressed in different ways for different purposes: as *fluence* (quantity per area), as a *rate* (quantity per time) or as *fluence rate* (quantity per area per time). Examples of the use of these different expressions are given in Table 2.2.

Conversion between measurements of energy fluence rate and quantum (photon) fluence rate is relatively straightforward, as long as the wavelengths involved are known; if it is a spectral band which is being considered, the middle wavelength can be used for purposes of calculation. The relationship between **energy fluence rate** ($W\ m^{-2}$) and **photon fluence rate** (moles $m^{-2}s^{-1}$) is:

$$W\ m^{-2} = moles\ m^{-2}s^{-1} \times \frac{Nhc}{\lambda}$$

where N = Avogadro's number; h = Planck's constant; c = velocity of light; λ = wavelength, in metres; and Nhc = 0.1196.

Therefore:

$$W\ m^{-2} = moles\ m^{-2}s^{-1} \times \frac{0.1196}{\lambda}$$

and

$$moles\ m^{-2}s^{-1} = W\ m^{-2} \times \frac{\lambda}{0.1196}$$

(For example, to transform 1 $W\ m^{-2}$ of visible radiation:

$$middle\ wavelength = 550\ nm = 550 \times 10^{-9}\ metres$$

therefore

28

$$\text{moles m}^{-2}\text{s}^{-1} = \frac{1 \times 550 \times 10^{-9}}{0.1196} = 4.6 \times 10^{-6}$$

That is, 1 W m^{-2} of visible radiation $= 4.6 \text{ }\mu\text{moles m}^{-2}\text{s}^{-1}$. And further, in temperate latitudes, *total* energy fluence rate at mid-day in summer is about 800 W m^{-2}; that is, about 400 W m^{-2} of *visible* radiation; therefore, the photon fluence rate for PAR is about $1,800 \text{ }\mu\text{moles m}^{-2}\text{s}^{-1}$.)

2.4.2 Quality

The quality of light in an environment is analysed by a **spectro-radiometer**. In this instrument, a scanning monochromator (e.g. a prism or diffraction grating plus filters) continually surveys the spectrum and disperses the light into its monochromatic components for individual detection and quantification by some kind of photo-sensor. The distribution of radiant energies in the spectrum is recorded on a chart (Fig. 3.2) or on some form of print-out. When the spectrum of any environment is being considered, however, it should always be borne in mind that photochemical processes arise from quantum events. Thus, the photon spectral distribution (numbers of photons per wave region) is more meaningful than the energy spectral distribution (joules per wave region). *'Light' is more properly described for photo-biological processes by photons per second per wavelength than by watts per wavelength.* (Some other aspects of light technology are discussed in Box 2.2.)

2.5 THE LIGHT SYSTEM

Radiant energy can only be observed by reason of its interactions with matter. This statement gives emphasis to the fact that the word 'light' always denotes a system made up of three components: a source of radiant energy; a medium through which the radiant energy passes; and a sensor which receives the radiant energy. These three components of the system persist, whether the effects of natural daylight on biological materials are being considered, or whether radiant energy from an artificial source is being measured. Unless all three components are kept under review, errors may arise in the comprehension of a situation. Some examples may illustrate this:

Source. Each source has its own specific photon spectral distribution. If a source has a relatively low photon output in the blue region

Box 2.2 Light filters

In experimentation with radiant energy, various types of light filter are frequently used (Holmes 1984a). *Neutral density* filters are non-selective across the spectrum and are used to reduce the quantity of irradiation. Commercial types have specified levels of transmission and can be combined to give the required reduction in photon fluence rate. Similar, but unquantified, effects can be achieved by layers of material like muslin or tissue paper.

With artificial sources of radiant energy, heat and infra-red radiation can pose problems. Some form of water-bath screen can be used to reduce heating effects, and a solution of copper sulphate can provide a cheap means of removing unwanted long wavelength radiation (Fig. 2.6).

Often, a specific region of the spectrum is required; filters for this purpose are of two general classes, selective absorption and optical filters. *Selective absorption filters* transmit wavebands which are not reflected or absorbed by materials in the filters. Such materials can be organic dyes, inorganic salts, or ions, carried in a medium of glass, plastic or gelatin. Although these filters can be large in size, they transmit relatively wide waveband regions and usually must be used in particular combinations. Since such filters are rarely manufactured specifically for investigations of plant responses, care must be taken to ensure that no unsuspected regions of the spectrum, particularly in the invisible far-red, are being transmitted.

With optical filters, laws of optics other than absorption – for example, refraction, reflection and interference – are used to control the transmission properties. Interference phenomena arise from the superimposition of some waves on each other, while other waves are out of phase; particular waves are thus reinforced or cancelled out. *Interference filters* consist of two semi-transparent layers whose spacing causes multiple reflection and constructive interference of the required wavelengths which thus pass through the filter; the other wavelengths interfere destructively, and are reflected back out of the filter. For example, in the Fabry–Perot type, the distance between the reflective layers is half the wavelength of the transmitted radiation. Such filters are highly precise in their transmission but they are limited in size, because of the requirements for optical glass and for control in layer deposition during manufacture. (Further, for precise transmission, such filters must be exactly perpendicular to a collimated light beam, otherwise the distance between the layers is not as specified.)

(e.g. an incandescent bulb), then it cannot be used to compare the relative physiological effectiveness of the blue and red regions of the spectrum, simply by inserting appropriate filters (Fig. 2.7a).

Medium. The specific spectrum of the source may be considerably altered by the particular absorption–transmission properties of the medium through which the radiation passes. In the natural

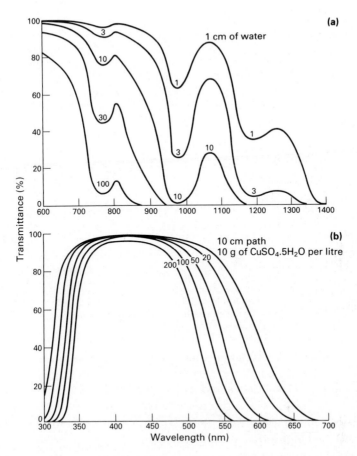

Figure 2.6 Spectral transmissions of water and of solutions of copper sulphate. (*a*) Water; the numbers on the curves indicate the path length in centimetres. (*b*) Copper sulphate solutions in a 10-cm path length at the indicated concentrations in 0.5% sulphuric acid.

environment, the atmosphere, vegetation cover and water act as selective filters on the daylight spectrum. The use of artificial filters can also create complications of a quantitative and qualitative nature. Quantitatively, to continue the example of the blue vs. red investigation, if the blue filter has a lower percentage transmission than the red filter, the original difficulties from the source would be exaggerated (Fig. 2.7b(i)). Qualitatively, if either filter transmitted invisible radiation in the far-red region of the spectrum, plant responses could be affected (Fig. 2.7b(ii)).

Sensor.   The receiver, whether a biological photoreceptor or some kind of measuring device, has its own specific sensitivity and absorption characteristics. If a photoreceptor cannot absorb the

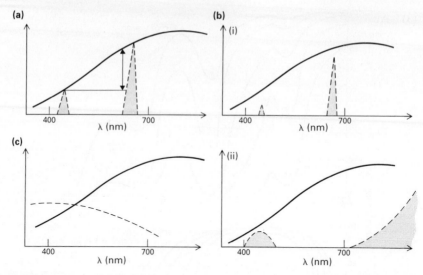

Figure 2.7 Examples of differences in 'source–medium–sensor' relationships in a light system: (*a*) source: a spectrum has less blue than red light; (*b*) medium: (*i*) a red filter has greater transmission than a blue filter; (*ii*) a 'blue' filter transmits far-red radiation; (*c*) sensor: sensitivity greater in the blue region of the spectrum.

available radiant energy, then, from the Grotthus–Draper first law of photochemistry, no biological effect can accrue. Again, in the example of the blue vs. red experiment, if the photocell of the light-measuring apparatus is more sensitive to the blue region of the spectrum, it cannot be used to equilibrate the irradiation treatments (Fig. 2.7c).

Therefore, the three components of a light system must always be considered, separately and together, whether a natural light environment is being investigated or whether a laboratory experiment is being devised.

2.6 SUMMARY

(1) Radiant energy exists in the continuum of the electromagnetic spectrum; it is propagated as waves, but interacts with matter as particles. The energy of a particle is inversely proportional to its wavelength.

(2) Radiant energy is emitted as a consequence of various processes. Thermal sources are characterized by continuous spectra whose shapes are determined by the temperature of the emitting substance; discharge sources emit line spectra.

(3) Absorption and emission of radiant energy are quantum phenomena, and represent specific changes of energy state in a molecule; molecules have characteristic absorption (and emission) spectra.

(4) When a molecule absorbs radiant energy, it is raised to a higher energy level and is more chemically reactive. Photon energies determine whether this can happen; photon numbers determine the extent to which it happens.

(5) The interaction of radiant energy with matter is formalized in the laws of photochemistry. The Fifth Law of Photochemical Equivalence states that one photon excites one electron.

(6) The excitation energy of a photoexcited molecule can be dissipated in a number of ways, which include thermal decay, fluorescence, resonance transfer and photochemical reaction.

(7) Light quantity can be measured photometrically, radiometrically and in quantum terms; photometric methods are inappropriate for investigation of plant processes; light is properly described for photobiological purposes as photons per second per wavelength.

(8) The term 'light' encompasses the complete system of source–medium–sensor.

FURTHER READING

Bickford, E. D. & S. Dunn 1972. *Lighting for plant growth*. Ohio: The Kent State University Press.

Clayton, R. K. 1970. *Light and living matter: a guide to the study of photobiology*. Volume 1: The physical part. New York: McGraw-Hill.

Seliger, H. H. & W. D. McElroy 1965. *Light: physical and biological action*. New York: Academic Press.

CHAPTER THREE

Daylight and artificial light

3.1 THE SUN'S SPECTRUM AND DAYLIGHT[1]

The Sun is a nuclear-powered, thermal source of radiant energy, and emits a continuous spectrum. The amounts of radiation in the different regions of this spectrum before it passes through the Earth's atmosphere are shown in Table 3.1. Obviously, the greater part of the spectrum is described by wavelengths longer than 400 nm. Significant amounts of energy are not found beyond λ 1500 nm, although the spectrum does extend to much longer wavelengths (radio-waves, i.e. very long wavelengths of en radiation, from the Sun were first detected in the Sun's spectrum in the 1940s).

Marked changes in the spectral distribution occur by the selective actions of various factors during passage through the atmosphere

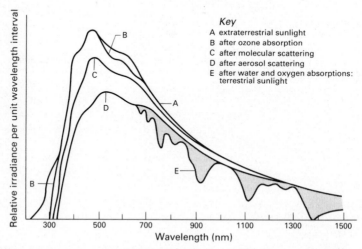

Figure 3.1 The spectrum of sunlight, progressively modified by various processes as it penetrates the atmosphere.

[1] See Giese (1978), Seliger & McElroy (1965), Smith (1982).

Box 3.1 Ultraviolet radiation

Radiant energies of $\lambda\ 40 - \lambda\ 400$ nm are classified as various types of ultraviolet radiation. The classes of UV are somewhat arbitrary, and are not always described by exactly the same limits in different texts (see Giese 1978).

Vacuum UV ($\lambda\ 40 - \lambda\ 200$ nm). The term 'vacuum UV' derives from the fact that this type of radiation cannot persist in the presence of oxygen. The oxygen is ionized to ozone, which absorbs all energies of less than $\lambda\ 200$ nm.

Far UV ($\lambda\ 200 - \lambda\ 300$ nm). This is the region which has lethal effects on biological materials (and also sun-tanning properties). It is subdivided into UVC ($\lambda\ 200 - \lambda\ 280$ nm) and UVB ($\lambda\ 280 - \lambda\ 320$ nm).

Near UV ($\lambda\ 300 - \lambda\ 400$ nm). This region approximates to UVA ($\lambda\ 320 - \lambda\ 390$ nm). Radiation of around $\lambda\ 380$ nm is the limit which can be detected by the average human eye. (The ageing process tends to raise this limit to longer wavelengths. Conversely, surgical removal of cataract often results in the ability to detect shorter wavelengths.)

The lethal and bactericidal actions of UV radiation are due to its deleterious effects on nucleic acids, rather than to effects on cytoplasm or proteins. Photoexcitation induces dimerization between nucleic acid bases, which prevents subsequent replication and expression of the genetic material.

Ultraviolet radiation can have photomorphogenic effects on plants. Most of the blue-light receptors (see Ch. 4) are sensitive to near-UV ($\lambda\ 370$ nm). A few responses to far-UV have also been noted, such as anthocyanin synthesis in sorghum, and phototropism in the grass coleoptile ($\lambda\ 270$ nm). Present opinion tends to the view that UV radiation in the natural environment has no great influence on plants, and certainly no deleterious effects, even in habitats where its level is somewhat higher than normal (e.g. at altitude). Plants in such habitats have presumably acquired the relevant protective features (high flavonoid pigmentation and repair mechanisms) (see Wellman 1983).

In certain situations, UV radiation can have direct effects on plant growth. In greenhouses, stray radiation from old or damaged lamps can carry a high proportion of UV wavelengths (plant symptoms range in severity from the development of a very thick cuticle, through distortion of the leaves to actual necrosis). Again, seedlings can become 'sunburnt' if planted out suddenly in a UV-enriched location (foresters minimize this process of solarization when planting at any altitude by placing tree seedlings in protected situations). Ultraviolet radiation also has many indirect effects on plant growth, through its ionizing and oxidizing actions on atmospheric pollutants; both smog and acid rain result from UV-induced reactions.

Table 3.1 Spectral distribution of sunlight incident on the atmosphere (adapted from Seliger & McElroy 1965).

Wavelength (nm)	% energy	W m^{-2}	mol s^{-1} × 10^8 over total surface
below 200	0.1	1.36	2.5
200–300	3.0	40.8	114
300–400	8.9	121	450
400–700	36	490	2,880
700–1000	24	326	2,950
above 1000	28	381	—

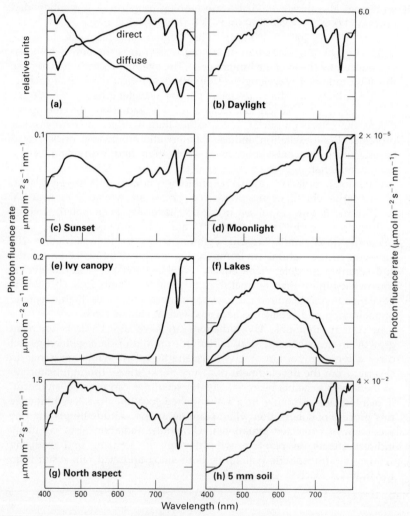

Figure 3.2 Typical spectroradiometer scans of the 400 nm to 800 nm waveband in different natural light environments.

(Fig. 3.1). Ultraviolet (UV) radiation is absorbed by atmospheric ozone, so that the cut-off point for UV radiation in direct sunlight is just below λ 300 nm (Fig. 3.1, curve *B*). If a quartz diffuser is used on the measuring instrument, then wavelengths of 285 nm can be detected. (The UV region of the spectrum is discussed further in Box 3.1.) The visible region is considerably modified by atmospheric scattering effects. These include Rayleigh (or molecular) scattering, where the wavelength of the radiation is larger than the interacting molecule (curve *C*), and Mie (or aerosol) scattering, where the wavelength is shorter than the diameter of the scattering particle (curve *D*). These effects particularly attenuate the shorter (blue) wavelengths. Therefore, the greater the air mass through which sunlight passes, the more its spectrum is weakened in the blue region. For example, when the Sun is directly overhead, the peak energy emission is around λ 500 nm; when it is within 10° of the horizon, peak emission is around λ 600 nm. The spectrum is further modified in the near infra-red through absorption of .specific regions by water, oxygen and carbon dioxide. Curve *E* (Fig. 3.1) represents the spectrum of sunlight incident on the Earth's surface.

'Daylight', however, is made up of two components: *direct sunlight* (or incident-light or I-light), with a spectral distribution approaching curve *E* (Fig. 3.1); and *skylight* (or diffused light or D-light), which is relatively enriched in the blue region of the spectrum due to particle scattering effects (Fig. 3.2). The proportions of sunlight and skylight in the total spectrum vary according to conditions: under cloudless skies, sunlight can constitute 80% of daylight; under completely overcast cloud, or when the Sun is below the horizon, daylight can be virtually 100% skylight. This daylight spectrum is usually subject to much further modification.

3.2   FACTORS WHICH INFLUENCE THE DAYLIGHT SPECTRUM[2]

The daylight spectrum can be so substantially modified by such a variety of factors that distinct spectral environments exist in different habitats. This aspect is only beginning to be investigated by plant scientists. (In the past, physicists and meteorologists have generally been concerned with spectral variation on a scale larger than is relevant to any individual plant.) The spectral regions of most significance to plant development consist of blue, red and far-red radiation. In fact, the relative amounts of red and far-red radiation, termed the R : Fr ratio

[2] See Holmes (1981), Smith (1982, 1983b), Smith & Morgan (1983), Spence (1981).

Table 3.2 Factors which modify the daylight spectrum.

Factor		Property of light environment affected
elevation	topography	mainly quantity
	time	quantity and quality
	latitude	quantity, quality, duration, periodicity
cover	cloud	mainly quantity
	vegetation	quantity and quality (R : Fr < 1)
	water	quantity and quality (R : Fr > 1)

or ζ (zeta), have assumed particular importance because of effects on the photoequilibrium and biological activity of phytochrome (Ch. 4).

Factors which modify daylight can be descriptively classified under two general headings: elevation of the source (i.e. angle between the source and the receiver); and cover (i.e. filter between the source and the receiver). A summary of the effects of these factors on the daylight spectrum is given in Table 3.2.

Elevation

Change in elevation of the Sun has both qualitative and quantitative effects on the spectrum. Qualitative effects result from the greater or lesser air mass through which the radiant energy passes, with consequent absorption and scattering of particular spectral regions. Quantitative effects arise because, if radiation impinges on a surface at angles of incidence other than normal to the plane of irradiation, it will be distributed over a greater area (Fig. 3.3); the fluence rate will thus be

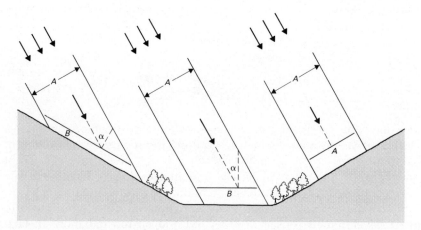

Figure 3.3 The Lambert cosine law: fluence rate changes according to the cosine of the angle of incidence. Angle of incidence (α) varies with orientation of the receiving surface as well as with the position of the light source. Fluence rate 'B' = fluence rate 'A'.cosα.

lower. The Lambert cosine law formalizes this relationship in stating that the irradiation of a surface varies with the cosine of the angle of incidence. (The angle of incidence is the angle between the normal to the surface and the direction of irradiation.)

Changes in the elevation of the Sun relative to the receiver can be brought about by a variety of agencies, including topography, time (of day and year) and latitude. Effects of *topography* are largely quantitative; there can be a 20-fold difference in fluence rate between a south-facing and north-facing slope. Differences in quality due to greater air passage or to a higher content of scattered skylight in the spectrum are probably negligible in relation to such large quantitative differences.

Effects of elevation arising from differences in *time of day* are both quantitative and qualitative. The photon fluence rate ranges from around 1800 µmole $m^{-2}s^{-1}$ in daylight to 0.005 µmole $m^{-2}s^{-1}$ in moonlight. Striking changes also occur in the spectral distribution. As the Sun declines towards the horizon, the relative amounts of blue radiation decrease, due to scattering and absorption by a greater air mass: with continued declination, the amounts of red also decrease until, when the Sun is 10° above the horizon (twilight), the R : Fr ratio is around 0.8. However, at the point of sunset, a transient increase in the relative amount of blue radiation occurs (Fig. 3.4), due to the high proportion of skylight in the spectrum at this stage. At the start of the daily cycle, similar changes occur in reverse order. Thus, beginning and end-of-day are characterized by transient blue-enhancement and lowered but changing R : Fr ratios.

Effects of *latitude* on the angle between the Sun and the receiver are

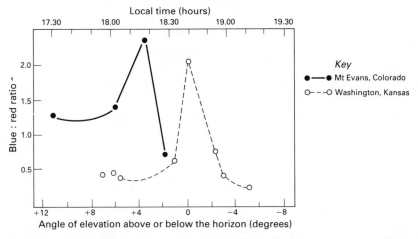

Figure 3.4 Changes in the blue : red ratio of daylight, during sunset at two different locations in the USA. The angle in degrees above the horizon refers to the upper limb of the Sun.

complex, and involve interactions with other factors, such as time of year, climate and so forth. Quantitative differences between, for example, the Tropics (2000 µmole $m^{-2}s^{-1}$) and temperate latitudes (1800 µmole $m^{-2}s^{-1}$) are not large in themselves, but cloud cover exerts a further massive influence in temperate zones. The quality of the spectral environment also differs between low and high latitudes: at high latitudes, the daily cycle of spectral change is not as extreme (the Sun does not rise as high in the sky); but equally, at latitudes where the Sun never rises much above the horizon, plants must exist in a light quality approaching that of twilight. Possible effects from such features have not yet been fully investigated. However, latitude does interact with one aspect of the light environment that has obvious effects on plant development, namely, periodicity of irradiation: with increasing latitude, seasonal differences in daylength markedly increase (Fig. 3.5).

Cover

The spectra of different habitats are subject to much further modification by the filtering actions of such agencies as cloud, vegetation canopies and water. The effects of *cloud* are mainly quantitative – that is, cloud acts as a neutral density filter. The photon fluence rate can be reduced

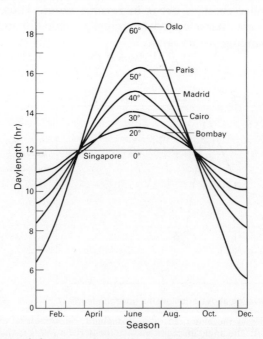

Figure 3.5 Seasonal changes in daylength at different latitudes.

Table 3.3 Ratios of R : Fr radiation in various environments.

Light type or environment		R : Fr ratio
natural light	daylight	1.19
	twilight	0.7–0.9
artificial light	incandescent	0.70
	fluorescent	13.5
water, 1 m deep	peaty	17.2[a]
	calcareous[a]	1.2[a]
vegetation canopies	wheat	0.2–0.5
	sugar beet	0.03–0.04
	deciduous forest (beech)	0.36–0.9
	coniferous forest	0.15–0.76
	tropical forest	0.22–0.77
soil, 5 mm deep		0.5–0.8

Note: Ratios compiled from: Holmes (1981), Smith (1982), Smith & Morgan (1983) and references therein.
[a]Data from Spence (1981), ratios computed by Smith (1982).

40-fold under an overcast sky. In comparison to such large quantitative effects, the slight relative enrichment in the blue region of the spectrum, due to the increased proportion of skylight, is probably insignificant. Similarly, the qualitative effects of atmospheric pollutants, in attenuating the blue region by Mie scattering, are probably much less significant than their chemical effects.

Vegetation is the most significant filter of the radiation received by terrestrial plants. Canopies have large effects on both quality and quantity of light. Under a canopy, the photon fluence rate is greatly reduced (Fig. 3.2). Since this is due mainly to absorption by photosynthetic pigments, the R : Fr ratio is also altered (Table 3.3). Thus, in such an environment, there are features that could enable a plant to distinguish between being shaded by other vegetation and by inanimate objects. (In the latter situation, quantity would be lowered, but normally there would be no great change in spectral distribution.) Spectral changes caused by canopy shade are also different from effects due to dawn and dusk (where the relative levels of the blue region of the spectrum and the R : Fr ratios are in a continual state of change). There are indications that particular types of canopy shade have their own characteristic features: the spectrum under broadleaf deciduous canopy tends to have a minor peak in the green region (Fig. 3.2e), while that of a coniferous canopy shows a small peak in the blue region. Again, different canopy types have characteristic R : Fr ratios (Table 3.3). However, this is a complex area which is not yet wholly resolved. The spread of values for the R : Fr ratios of particular canopies (Table 3.3), is related not only to details of canopy structure, age and density, but also to the interaction of other factors which influence the light environment. For example, cloud cover, while not itself greatly affecting spectral quality, actually lessens the effects of vegetation canopy; the more diffuse light under cloud enters the canopy at many different angles. Another aspect of canopy light is discussed in Box 3.2.

Box 3.2   Sunflecks

The spectral environment in a canopy is further complicated, quantitatively and qualitatively, by sunflecks where I-light (see text) enters through a gap in the canopy (Holmes 1981). Sunflecks vary in *size* and the related aspects of *duration* and *periodicity*. For example, 'windflecks' can flutter between conditions of sun and shade at around 20 cycles s^{-1}. On the other hand, movement of the Sun across larger gaps in the canopy would result in quite a different pattern of exposure: a gap that is roughly equivalent in size to the disc of the Sun itself when viewed through the gap would expose some ground positions within it to sunlight for periods of up to five minutes; larger gaps, of course, would give even longer exposures. (Such sunflecks that are related to the movement of the Sun have been referred to as 'timeflecks'.)

The general effects of sunflecks on plant growth are therefore problematical. With regard to photosynthesis, there are indications that although sunflecks represent substantial increases in levels of available energy, the plant cannot always make use of them. The photosynthetic machinery that is developed under shade conditions, consisting of characteristic leaves and types of photosynthetic units, cannot process greatly increased levels of radiant energy. In relation to photomorphogenic effects, the nature and extent of any response to sunflecks will depend in the first place upon whether a 'shade-avoiding' or 'shade-tolerating' type of plant is being considered (see Ch. 6). Secondly, it is likely that a plant ignores the spectral 'noise' of windflecks, certainly as far as phytochrome-mediated responses are concerned: it takes a finite time for the phytochrome photoequilibrium (or level of 'active phytochrome', see Ch. 4) to be established (e.g. one minute at 10 Wm^{-2}). However, 'timeflecks' of 10 minutes or more could have significant effects: phytochrome-mediated growth responses have been detected within this period, and responses to blue light are even more rapid (Ch. 6).

Actual responses to sunflecks in the natural environment have not yet received much study. In the case of *Oxalis acetosella* (wood sorrel), exposure to natural sunflecks results in leaf closure. That is, this plant, whose natural habitat is deep shade, actually protects itself from the potentially damaging light intensities of sunflecks. This is a photonastic response mediated by blue light and it takes place within three minutes.

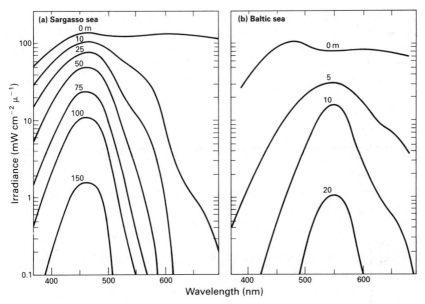

Figure 3.6 Spectral distributions of downward irradiation at different water depths in marine environments that are (a) relatively clear or (b) polluted.

Water, and the substances in it, cause substantial change in the spectral distributions within marine environments. Again, the effects are complex and are influenced by such factors as angle of light penetration, depth, amount and type of solutes and particulate matter. The exponential decrease in photon fluence rate with depth is affected also by the level of 'solute', both chemical and biological (Fig. 3.6); the photosynthetic compensation depth varies from 10 m, in a highly polluted environment such as the Baltic Sea, to 150 m in the unpolluted Sargasso. Similarly, changes in the spectral quality depend for their detail on the particular marine environment. On the one hand, the longer wavelengths are progressively attenuated by increasing depth; on the other hand, the shorter wavelengths are attenuated through absorption and scattering effects by the solutes. Plant material removes the blue and red regions of the spectrum. A unique feature of aquatic environments is the increase in the R : Fr ratio to values much greater than 1.0 (Table 3.3).

3.3   ARTIFICIAL SOURCES OF RADIANT ENERGY[3]

Man's behaviour is no longer governed by the periodicity of sunlight – artificial light has allowed his activities to extend into a

[3] See Bickford & Dunn 1972, Holmes (1984a), Vince-Prue & Canham (1983).

considerable range of times and places. However, it should be borne in mind that such lights have almost invariably been developed to aid human vision, and are thus not always suitable for the requirements of plants. (The problem can be illustrated by noting the results when a plant with fairly exacting light requirements is introduced into a 'well lit' domestic environment – even the common geranium only survives in a rather etiolated condition in most business premises). Emission of radiation in response to thermal energy and by electric discharge are the main methods used to generate light, with re-emission of radiation as fluorescence also playing a major role. The types of lamp in common use and their chief characteristics are summarized in Table 3.4. In the description of an artificial light source, the important features are the fluence rate and the spectral distribution. The 'efficiency' of the source is a term used by lighting engineers and refers to the *visible* brightness per energy input (lumens per watt). It is therefore not directly relevant to considerations of plant growth, but it does give a general indication of the potential fluence rate.

Thermal sources

A candle flame represents one of the most primitive means of obtaining radiant energy through thermal emission. An **incandescent** light source operates by the same principle, except that electrical energy is used to heat the filament. The distribution of the continuous spectrum from such a source depends on the temperature of the filament, but all such lamps are characterized by emission of large amounts of infra-red radiation and relatively small amounts of light (Fig. 3.7). The filament is usually *tungsten*, which has a high melting point; at greater than 40 watts, an inert gas is enclosed in the bulb to minimize tungsten evaporation. The temperatures at which such sources operate are still relatively low (filament 2000°C, bulb surface 200°C), giving low brightness and an efficiency of around 7%. Eventually the tungsten does evaporate (bulb blackening), giving this type a relatively short life.

The *tungsten–halogen* bulb (old name is 'quartz–iodide') contains bromine vapour to protect the filament; a tungsten × bromide chemical cycle redeposits tungsten on the filament. These bulbs therefore not only have a longer life, but operate at higher temperatures. The spectrum is shifted slightly towards the shorter wavelengths – that is, the lamps are 'brighter' (efficiency, 10%). (Note that attempts to vary the fluence rate from these sources by change in voltage supply will also result in significant change in spectral distribution.) The higher operating temperatures of these lamps mean that the envelope must be quartz rather than glass (the bulb should not be handled directly since

Table 3.4 Characteristics of common artificial sources of radiant energy.

Type of mechanism		Type of lamp		Characteristics
thermal	incandescent	tungsten		continuous spectrum, low PAR, low blue, high Fr, high heat
		tungsten-halogen		higher temperatures, brighter
discharge	luminescent	Hg	low pressure	mainly UV (254 nm), bactericidal
		Hg	high pressure	high fluence rate, major lines at 365, 405, 436, 546, 578 nm + phosphor = continuous background emission + metal halide = more emission lines in the red
		Na	low pressure	monochromatic (589 nm), high fluence rate
		Na	high pressure	high fluence rate at 550–620 nm
		Xe	high pressure	high fluence rates, good spectral match with daylight
	fluorescent	Hg	low pressure	various spectra, low fluence rates, low Fr

sweat deposition will lead to breakage through the generation of local heat-spots). The heat from such sources is considerable, and usually they must be used in conjunction with a water-bath filter or a dichroic reflector.

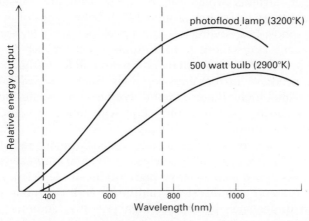

Figure 3.7 Relative spectral energy distributions of two incandescent lamps operating at different temperatures; the visible region is indicated by dashed lines.

Thus, incandescent sources of radiant energy are characterized by continuous spectra which are low in the blue region and high in far-red wavelengths; they are in general inefficient, with low fluence rates and short life spans. They are cheap, but biologically poor.

Electric discharge sources

In discharge sources, electrons are accelerated off a cathode to excite particles of a gas which emit line spectra characteristic of the type of gas and its pressure. Since they operate on a lower energy input, they are more efficient than incandescent sources, and are capable of providing relatively high fluence rates in certain parts of the visible region. There are two types of discharge source: in **luminescent** sources the radiant energy emitted by the excited gas molecules is used directly as light; in **fluorescent** lamps, the primary emission is used to excite a second class of molecule (the phosphor), which in turn emits secondary (fluorescent) radiation in the visible region.

Luminescent
Different types of luminescent lamp contain different gaseous phases at different pressures.

Mercury vapour, at low pressure (around 0.05% atm.) emits primarily UV radiation at λ 253.7 nm (Fig. 3.8); when enclosed in a quartz envelope, such a source acts as a bactericidal lamp. With increasing pressure, the other five major mercury emission lines (λs 365, 405, 436, 546, 578 nm) assume greater prominence (Fig. 3.8). At even higher pressures (up to several atmospheres), the emission lines are broadened and the higher currents which can be used result in considerable increase in brightness (efficiency around 14%). Some types of mercury lamp carry a phosphor coating on the envelope; such *high-pressure mercury fluorescent* lamps emit a continuous spectrum which peaks around λ 660 nm. Addition of various rare earth halides to the mercury vapour phase produces further emission lines. Such *metal halide* lamps therefore provide high fluence rates coupled to some spectral flexibility; they can be used in conjunction with filters as sources for monochromatic radiation.

Sodium vapour at low pressure itself emits monochromatic radiation at λ 589 nm (Fig. 3.8). Because this is so close to the maximum sensitivity of the human eye, such sources have high luminosity (efficiency 32%, hence their common use in street lighting – and the poor colour discrimination under such lights). At high pressure, sodium vapour sources emit many more lines in the orange–red region, λs 550–620 nm (Fig. 3.8). Therefore, high-pressure sodium lamps are used as relatively

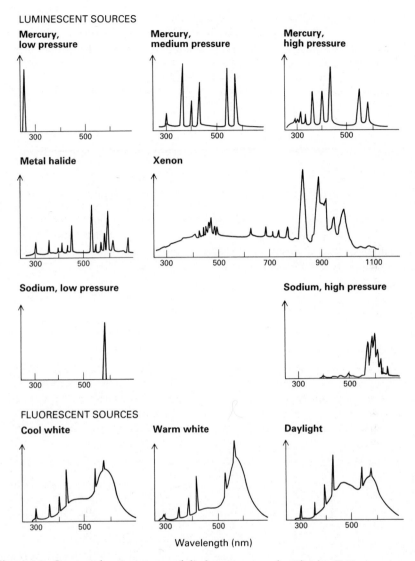

Figure 3.8 Spectra of various types of discharge source of artificial radiation.

cheap sources of supplementary radiation for photosynthesis.

Xenon vapour emits a great many emission lines. Xenon or mercury–xenon lamps therefore provide high fluence rates in a very good spectral match with sunlight (Fig. 3.8). (Installation and running costs are, however, high.)

Fluorescent

Fluorescent lamps are all based upon mercury vapour, at a variety of

Box 3.3 Artificial lighting for plant growth

In temperate zones, daylight can be limited in quantity or duration and some form of artificial lighting for plant growth is often required (Vince-Prue & Canham 1983). All considerations of such systems revolve around the three elements of: *cost, purpose, return*. Aspects of cost and return are relatively straightforward. Costs include those for installation, operation and renewal. Return values, whether of experimental results or of a commercial product, depend either upon a high yield or a certain rarity effect; as commercial examples, either a high density of seedlings per irradiated area or floral blooms out of season. For these aspects, specific photobiological knowledge need not be a high priority. However, considerations of purpose do require some photobiological awareness.

There are two quite different ways of using artificial light: wholly as a replacement for natural light or merely as a supplement. There are two quite different types of process for which plants require light: photosynthesis and photomorphogenesis. The table shows the simple permutation of these situations:

	Replacement	*Supplement*
Photosynthesis	(1)	(2)
Photomorphogenesis	(3)	(4)

Photosynthesis
(1) In general, at today's market prices, photosynthetic production wholly under artificial light is not financially viable. (It is, however, perfectly feasible technically, and this 'market decision' can change with circumstances, e.g. in Siberia, although equally nutritious vegetables could be imported into the region, plants are grown wholly under artificial lighting in order to put some 'colour' into local life.)
(2) If daylight is being supplemented by artificial light for photosynthesis, then the fluence rate of the artificial light is more critical than its spectral distribution. The efficient, high-pressure, sodium discharge lamps can suffice for this, although metal halide–Hg sources offer greater spectral flexibility.

Photomorphogenesis
If the artificial light is used just as a signal input, then the spectral distribution is more important than the fluence rate; duration and timing of irradiation also need to be considered.
(3) If the artificial source is wholly to replace daylight, then only some form of non-photosynthetic development can be financially considered, i.e. the plant material must utilize its own reserves. Artificial light is used in this way for rhubarb growth, seedling production, potato sprouting and some types of flower production. For this, the low cost, low fluence rate systems of fluorescent combined with incandescent sources are satisfactory.
(4) The major commercial use of artificial light is as supplementary light

signals for growth under natural light, particularly for the control of photoperiodically sensitive flowering. Again, the low-cost systems comprising fluorescent and incandescent sources are suitable. (Note that it is technically more simple to extend daylength by the application of a light signal than it is to shorten the day by the total exclusion of light; and it is cheaper to extend the day by early morning light than by irradiation in the evening.) Photoperiodic control of flowering serves a variety of 'market' purposes. It can be used to control the *timing* of flowering for commercial reasons (e.g. Christmas flowering poinsettias are produced by short-day regimes, and winter carnations (a 'quantitative' long-day plant, see Ch. 8) by 2–3 weeks of continuous light). Similarly, it can be used to control the *form* of a plant for commercial or aesthetic reasons – for example, small flowering pot plants of chrysanthemum (a short-day plant – see Ch. 8) are produced by placing cuttings in regimes of short days soon after they have rooted, but larger plants that are more suitable for the cut flower industry are produced by growing the rooted cuttings under long days for a period. The subsequent development of flowers in chrysanthemum is also enhanced by short days (as well as initiation of flowering), and so particular regimes of short days and long days can control the form of the actual floral sprays through effects on flower growth and pedicel growth.

pressures. Primary emission is therefore in the UV region of the spectrum, with greater or lesser amounts of the other five mercury emission lines. A phosphor coating on the lamp envelope absorbs this primary emission, and re-emits fluorescent radiation in the visible region. The phosphor is a mixture of zinc–beryllium silicate and magnesium tungstate in different ratios in lamps intended for different purposes. Spectra of fluorescent lights thus show the five mercury lines overlaid on a secondary continuum whose shape depends on the type of phosphor (Fig. 3.8). Examples of different types of fluorescent light for different purposes are *Warm White* for domestic use, with a lot of red light in it; *Daylight* for better visibility, with more blue in it; and the *Gro-lux* tube, developed by Sylvania for particular application to plant growth, with a high proportion of its output in the blue and red regions. It is thus clear that it is not enough merely to describe an irradiation source as 'fluorescent'. All fluorescent sources have a good efficiency (around 22%), a relatively long life, and are cheap to install and operate. However, they are capable of providing only limited fluence rates, and are characterized by having little or no radiation in the far-red region. This last aspect can be accommodated by including incandescent sources in an irradiation system (traditionally, in the wattage ratio of 3 fluorescent : 1 incandescent). The low fluence rate of the fluorescent source is compensated to some extent by its shape as a bar source of radiation. (The positioning of point sources, even with reflectors, must be considered carefully, due to the combined effects of the Lambert cosine law and fluence rate diminution according to the square of the distance from the source.) The output of fluorescent lights is markedly affected by their age, and the temperature at which they are operated; output can fall by as much as 30% in their first six weeks of use, before remaining stable; low temperatures, even of 15°C, can significantly reduce output through effects on the internal gas pressure. The use of artificial lighting for plant growth is discussed further in Box 3.3.

3.4 SUMMARY

(1) The Sun emits a continous spectrum of radiant energies; atmospheric effects modify the spectrum to permit the passage of a region from λ 300–1500 nm, with a relatively flat daylight photon spectral distribution between λ 400–700 nm. Daylight is made up of direct sunlight and skylight.

(2) The daylight spectrum is further modified by various situations, such as elevation of the source, which may be affected by topography, time and latitude; and type of cover (cloud, vegetation canopies, water etc.). Major qualitative changes in the spectrum are the low R : Fr ratios at dawn and dusk and in vegetational shade. Latitude markedly influences the periodicity of irradiation.

(3) Artificial lights are thermal or electric discharge sources of radiant energy. Incandescent lights have low visible fluence rates and high levels of infra-red radiation; luminescent lamps have high fluence rates and their own characteristic spectra; fluorescent sources have low fluence rates and little far-red radiation.

(4) The use of artificial light for plant growth tends towards one of four categories: (1) wholly to replace or (2) partly to supplement daylight in (3) photosynthesis or (4) photomorphogenesis. Photosynthetic production wholly under artificial light is in general not financially viable; if artificial light is supplementing daylight for photosynthesis, fluence rate is the important consideration; for control of photomorphogenic responses, spectral distribution is more critical than fluence rate.

FURTHER READING

Bickford, E. D. & S. Dunn 1972: *Lighting for plant growth*. Ohio: Kent State University Press.

Holmes, M. G. 1984. Light sources. In *Techniques in photomorphogenesis*, H. Smith & M. G. Holmes (eds.), 43–80. London: Academic Press.

Smith, H. 1982. Light quality, photoreception and plant strategy. *Annu. Plant Physiol.* **33**, 481–518.

Spence, D. H. N. 1981. Light quality and plant responses underwater. In *Plants and the daylight spectrum*, H. Smith (ed.), 245–76. London: Academic Press.

Vince-Prue, D. & A. E. Canham 1983. The horticultural significance of photomorphogenesis. In *Photomorphogenesis*, Encycl. Plant Physiol. NS 16B, W. Shropshire & H. Mohr (eds.), 518–44. Berlin: Springer.

CHAPTER FOUR

Photoreceptors

4.1 INTRODUCTORY COMMENTS[1]

Organisms contain many types of molecules that absorb radiant energy, ranging from nucleic acids and proteins, which interact predominantly with the ultraviolet region of the spectrum, to pigments which are coloured because they absorb certain visible wavelengths and reflect others. Higher plants exhibit a tremendous range of coloration, seen particularly in their flowers and autumn leaf displays. This range is achieved through the interaction of three chemical classes of pigment: the **chlorophylls** are responsible for the predominant green of nature; the **carotenoids** impart many of the strong yellow-orange colours; and the **flavonoids** contribute blues, purples and reds (through the anthocyanins) and yellow tones (through the flavones, flavonols, chalcones and aurones). Out of this variety of light-sensitive compounds, only a relatively few types of molecule have been selected to function as photoreceptors. The vast majority of pigments, although they absorb light, are not involved in the direct utilization of radiant energy. (For example, the flavonoids are, in the main, located in the vacuole of the cell, and function simply as visual attractants for pollination and seed dispersal by animals.)

Pigments which act as photoreceptors can be functionally grouped into two classes, photosynthetic and photomorphogenic. The structural formulae for the types of pigment involved in these processes are shown in Figure 4.1. The major photosynthetic pigments are, of course, the chlorophylls, which include the different chlorophylls a, b, c, d and bacteriochlorophyll. Chlorophyll a is common to all oxygen-evolving organisms; the content of other chlorophylls varies with the type of organism and the spectral environment. The other photo-synthetic pigments are termed accessory pigments, and include the carotenoids, which consist of the hydrocarbon carotenes and their oxygenated derivatives the xanthophylls, and the phycobilins, whose

[1] See Presti (1983), Presti & Delbruck (1978).

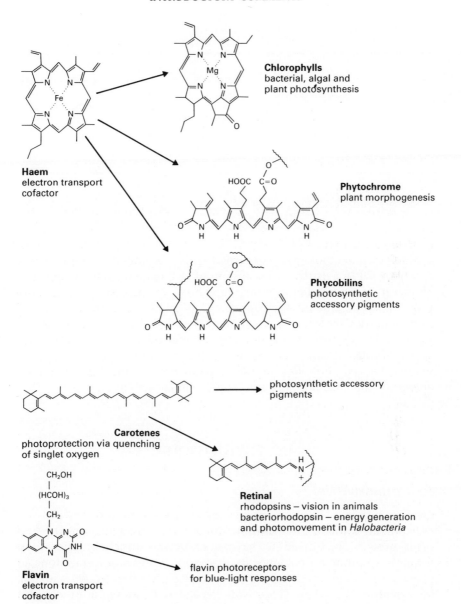

Figure 4.1 Basic structures, and possible evolutionary origins, of the various types of molecule which act as biological photoreceptors.

distribution is limited to certain marine algae. Although localized in a chloroplast in a higher plant, the photosynthetic pigments have a great influence on light absorption by the photomorphogenic pigments, through effects of screening and fluorescence.

Table 4.1 Germination of lettuce seed (var. Grand Rapids) after various sequences of red (R) and far-red (Fr) irradiation

Light treatment	Germination (%)
darkness	21
5 min R	92
5 min R + 10 min Fr	38
R–Fr–R	88
R–Fr–R–Fr	42
10 min Fr	40

The most clearly characterized photomorphogenic receptor is phyto-chrome, the non-protein part of which is structurally related to the phycobilins. Candidates for the as yet unidentified blue-light receptors include carotenoids and flavins.

Much of photobiology is concerned with the properties of pigments. (See Box 4.1 for further discussion of a pigment's characteristics). However, it should be borne in mind that, in the cell, each of these photosynthetic and photomorphogenic pigments is associated with a protein carrier molecule; the pigment is called the **chromophore** and the protein to which it is conjugated is referred to as the **apoprotein**. The apoprotein assists not only in the localization of the chromophore within the cell, but also in the transduction of the light-excitation energy of the chromophore into some cellular activity.

4.2 PHYTOCHROME

4.2.1 Introduction

Over a hundred different responses of plants to red and far-red radiation are known. Phytochrome, the receptor involved, was the first plant protein recognized to have a regulatory function other than through action as an enzyme. Its discovery represents a milestone in plant physiology, and serves as a classic illustration of scientific investigation: one key factor was recognized among a number of observations and led to a hypothesis that could be tested in several ways.

In 1937, the American investigators, Flint and McAlister, observed that red light promoted the germination of lettuce seed, while irradiation with far-red wavelengths inhibited it (Fig. 4.2). During the 1950s, in the US Department of Agriculture laboratories at Beltsville, a group led by Borthwick and Hendricks extended this finding and

Box 4.1 Pigments

One characteristic chemical feature of a pigment is a system of alternating double and single bonds (a conjugated double bond system, cf. Fig. 4.1). Electrons which participate in chemical bonds are distributed between the bonded atoms in certain ways. In a C–C single bond, the bonding electrons are localized symmetrically around the axis of the bond; this is described as a σ (sigma) bond. A double bond (C = C) contains one colinear bond and two parallel π (pi) bonds where the bonding electrons are not localized but are shared between the two atoms. In a conjugated double bond system, the electrons are further delocalized and shared between several atoms. It will be recalled that when a molecule absorbs radiant energy, its ground state is transformed to an excited state by the promotion of an electron to a 'higher' orbital ($hv = \Delta E$). A π orbital system of delocalized electrons has a high probability of entering an excited state – that is, a $\pi \rightarrow \pi^*$ transition requires a relatively low input of energy. (A $\sigma \rightarrow \pi^*$ transition has a lower probability and only occurs through absorption of the larger amounts of energy in UV radiation.) Furthermore, the greater the extent of the π orbital (i.e. the greater the delocalization), the smaller is the energy gap between the ground state and the excited state – the more double bonds there are in the conjugated system, the longer are the wavelengths of radiant energy which can bring about excitation. One double bond absorbs at λ 185 nm; a compound with two double bonds absorbs at λ 225 nm; a hydrocarbon chain with nine double bonds absorbs around λ 415 nm. (Note, too, that the last type of pigment absorbs in the blue region of the spectrum and is therefore yellow in colour.)

The absorption of light by molecules in dilute solution is described by the Beer–Lambert Law, which states that the fraction of incident radiation absorbed is proportional to the number of molecules in its path. The situation is formalized in the equation:

$$A = \log \frac{I/}{/I_0} = \varepsilon cl$$

where A = absorbance, I_0 = amount of incident light, I = amount of transmitted light, c = concentration of substance, and l = path length (cm) of light transmission through solution.

The term ε (epsilon) is the **absorption coefficient**, and is a function of the total absorbing system – the compound, its molecular environment and the wavelength of light. (The older term for this parameter is 'extinction coefficient'.) The absorption coefficient thus describes the capacity of a pigment in a certain solvent to absorb light at a particular wavelength: the more intensely coloured a pigment, the larger is its absorption coefficient. If the concentration of pigment is in moles per litre (dm^3), ε_λ is the *molar absorption coefficient* and is expressed in units of dm^3 mol^{-1} cm^{-1}. (Since A is a logarithm and has no units, the unit of ε_λ is simply the reciprocals of c, i.e. dm^3mol^{-1}, and l, i.e. cm^{-1}.)

Thus, a chromophore comprises an extensive conjugated double bond system, which absorbs broad wavebands in the visible region, and has a large absorption coefficient.

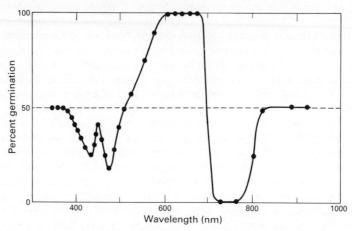

Figure 4.2 Action spectra for the promotion and inhibition of germination in light-sensitive lettuce seed.

showed that many other aspects of plant behaviour were affected by these regions of the spectrum. However, the key observation was that the effects of red light were *reversed* by exposure to far-red radiation, and vice versa; in fact, in most of the processes under investigation, if a sequence of alternating exposures to red and far-red radiation was provided, the response was determined by the final irradiation (Table 4.1). From this behaviour, it was postulated that the responses to red and far-red radiation were regulated by a single **photochromic receptor**, i.e. a receptor which changed its absorption properties after exposure to light. A special instrument, the dual-wavelength spectro-photometer, was developed to look for any such changes in absorption: this instrument can treat a sample with red or far-red radiation (these are called the 'actinic beams') while also measuring the absorbance of the sample at red and far-red wavelengths. The postulated absorption changes were found in (etiolated) tissues and in extracts: red irradiation gave an absorbance decrease at λ 660 nm and an increase at λ 730 nm; far-red irradiation resulted in an increase at λ 660 nm and a decrease at λ 730 nm (Fig. 4.3). (The method by which the relative amount of pigment can be determined from these changes is described in Box 4.2.)

Phytochrome is distributed throughout all classes of green plants, including algae, mosses, ferns and gymnosperms; it does not seem

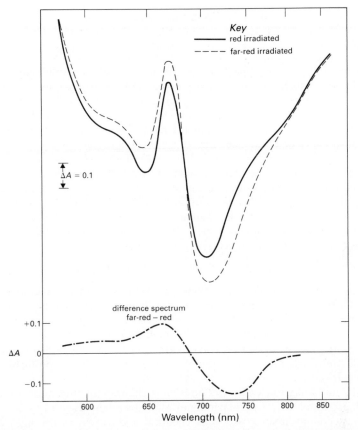

Figure 4.3 'Difference spectrum' for phytochrome *in vivo*. The upper curves are the absorbances after exposures to red and far-red irradiation; the difference between these curves is shown below.

generally to be present in fungi. It is a bluish chromoprotein (i.e. a chromophore conjugated to a protein moiety), and occurs in very low concentrations, around 2×10^{-7} mol dm^{-3} (2×10^{-7} molar) in etiolated oat seedlings, and even lower (2×10^{-9} mol dm^{-3}) in green seedlings. The chromophore exists in two forms, with accepted absorption maxima at λ 660 nm and λ 730 nm; both forms also show significant absorption in the blue region of the spectrum (Fig. 4.4). The photochromic reaction only occurs when the chromophore is conjugated to the apoprotein and is formally represented as:

$$P_r \xrightarrow[730 \text{ nm}]{660 \text{ nm}} P_{fr}$$

It is generally believed that P_{fr} is the biologically active form.

Box 4.2 Assay of phytochrome

A means of assay is a primary requirement for the investigation of any biological component. Spectrophometric techniques were originally used to detect phytochrome and, for many years, remained the only means of assay for it (see Pratt 1983). Spectrophotometric measurements of concentration in clear solutions are relatively straightforward, where absorption follows the Beer–Lambert Law, $A = \varepsilon cl$. (For phytochrome, an absorbance of 1.0 at λ 667 nm, in a 1.0 cm light path, is equivalent to 1.2 mg cm^{-3}.) However, when other interfering substances are present, the amounts of a photochromic pigment must be estimated from light-induced changes in absorption. The dual wavelength spectrophotometer measures absorption after red and far-red light treatments (Fig. 4.3). The amount of phytochrome is estimated by summing the differences in absorption at λ 660 nm and at λ 730 nm – that is, summing the two peaks of the lower difference spectrum of Figure 4.3:

$$(P = \Delta A_{660} + \Delta A_{730} = (A^{fr}_{660} - A^{r}_{660}) + (A^{r}_{730} - A^{fr}_{730})$$

There are two major limitations to spectrophotometric methods of phytochrome assay: only relative rather than absolute amounts can be determined; and susceptibility to interference from other pigments is such that it is impossible to use this method on green tissues. (Use of different pairs of measuring wavelengths, e.g. λ 660 nm : λ 800 nm and λ 730 nm : λ 800 nm, can minimize interference from protochlorophyllide).

The herbicide *Norflurazon* or San 9789 (4-chloro-5-(methylamino-2-$\alpha,\alpha,$-trifluoro-m-tolyl-3(2H)pyridazinone) inhibits carotenoid synthesis. Therefore, seedlings treated with this chemical and grown in the light are 'de-etiolated' but achlorophyllous, since their chlorophyll is no longer protected from photobleaching. Such non-green seedlings have been used for spectrophotometric assays of phytochrome in 'de-etiolated' (juvenile) tissues.

Recent developments in the field of phytochrome detection and assay involve the use of immunochemical methods. These provide quite different types of information from that obtainable by spectrophotometric assay since they are based upon properties of the apoprotein rather than the chromophore, and since they can be applied to the investigation of phytochrome in green tissues. Such studies have established that there is a great deal of structural similarity among phytochromes from a wide range of species, but they have also indicated that there are immuno-chemical differences between different phytochromes; these include differences between monocots and dicots, differences between the phytochromes from etiolated and green tissues of the same species, and even immunologically distinct forms of phytochrome within the same plant (see Lagarias 1985). The physiological significance of multiple forms of phytochrome has still to be assessed, but they are probably relevant to previously observed differences in spectrophotometric behaviour of

> phytochrome from different sources (e.g. from monocots and dicots); they may also be the basis of the different 'pools' of phytochrome that have been postulated to exist, from the results of certain physiological experiments.

Other methods by which the preliminary characteristizations of putative photoceptors can be made are described in Box 4.3.

4.2.2 Operational modes of phytochrome[2]

The traditional criterion for the identification of phytochrome action is the demonstration of R–Fr *reversibility* in a biological response. However, this simple statement requires some qualification:

(1) Reversibility operates only in responses induced by relatively brief periods of irradiation.

(2) Levels of (reversed) response with far-red irradiation are often higher than the level in darkness alone (Table 4.1). This results from the overlapping absorption spectra of the two forms of phytochrome (Fig. 4.4), and the consequent production of some P_{fr} by far-red irradiation itself. (The appropriate control treatment in reversibility tests is 'far-red alone', rather than darkness.)

(3) If broad band far-red radiation is used in reversibility tests, the far-red fluence usually must be about three times higher than that of the inductive red irradiation. This is largely due to $\varepsilon_{P_{fr}}$ and $\Phi_{P_{fr}}$ being lower than the corresponding values for P_r (and does not apply if λ 730 nm is used).

(4) Lack of apparent R–Fr reversibility in a response does not necessarily mean that phytochrome is not involved; very rapid action of phytochrome, or activity at very low concentrations of P_{fr}, can prevent demonstrable reversibility.

It should also be borne in mind that reversibility is a laboratory test for phytochrome; in nature, plants are not exposed to short periods of irradiation and phytochrome does not function as a simple on–off switch.

Besides R–Fr reversibility, the original criteria for phytochrome action included:

(1) reciprocity between duration of irradiation and fluence rate, up to saturating fluence levels;

[2] See Kendrick (1983), Mandoli & Briggs (1981), Mancinelli (1980), Mohr (1984), Schopfer (1984).

Box 4.3 Action spectra

An initial requirement in the investigation of any photoresponse is information on the nature of the photoreceptor. If the biological effect results simply from the excitation of a single pigment, then, as a direct consequence of the Grotthus–Draper Law, the wavelengths of greatest effectiveness must be those of maximum absorption. An **action spectrum** is a plot of relative sensitivity of a biological response to different wavelengths of irradiation. It is usually constructed for purposes of comparison with **absorption spectra** of known pigments. Any simple photoresponse (R) is a function of fluence rate (N), duration of exposure (t), absorption coefficient of the photoreceptor (ε) and quantum yield of the photoreaction (Φ):

$$R_\lambda \propto N_\lambda t \varepsilon_\lambda \Phi_\lambda$$

The most effective wavelengths will be those at which the *fewest* number of photons is required to produce the effect. However, if 'response' is plotted, against wavelength, as the *reciprocal* of the number of incident photons required to produce a given effect, then *peaks* in the resulting spectrum will represent the most effective wavelengths, and allow direct comparison with peaks in absorption spectra.

The first step in the determination of an action spectrum is the construction of *fluence-response curves* for different wavelengths (waveband regions) of irradiation. From these curves, the photon fluence required to produce a standard level of response is determined for each type of irradiation (i.e. the number of photons required to produce, say, 50% germination, or 50% inhibition of growth). The action spectrum is then prepared by plotting the reciprocals of these photon fluences against wavelength.

A good match between an action spectrum and the absorption spectrum of a putative receptor demands certain prerequisites, which are often difficult to fulfil in a biological situation:

(1) The absorption spectrum of the chromophore *in vivo* must be known. However, the absorption spectrum of a pigment is greatly influenced by its molecular environment – *in vitro*, by the type of solvent, *in vivo*, by association with other molecules. Therefore it is usually impossible to obtain an exact match between *in vitro* absorption and *in vivo* action. (Many determinations of absorption maxima for phytochrome *in vitro* do not coincide with λ 660 nm and λ 730 nm.)

(2) The fluence-response curves at different wavelengths must be similar in form (and preferably linear), to allow the conclusion that the photoreaction is similar. However, the presence of other pigments can distort these curves through effects of screening, scattering, and in the case of photosynthetic pigments, fluorescence.

(3) The extent of response must show reciprocity between exposure

period and fluence rate. This is expected in photochemical reactions, but reciprocity breakdown in biological responses can result from various situations, such as the involvement of more than one photoreceptor, the requirement for a threshold concentration of photoproduct, or the rapid degradation of photoproduct.

(4) The quantum yield for the response (Ch. 2, p. 23) should be independent of wavelength; if the quantum yield is not similar at all wavelengths, this could indicate various possibilities, from interference by screening pigments to the involvement of more than one photosystem. (The concept of quantum yield is difficult to apply strictly to biological situations, but some form of it is often useful for comparative purposes.)

Methods for dealing with many of these 'biological difficulties' are discussed in more advanced texts (see, for example, Schafër and Fukschansky 1984).

Action spectra have been used to great effect in the investigation of many photobiological responses, e.g. to demonstrate the involvement of carotenoids in vision, DNA in UV-induced mutagenesis and chlorophyll in photosynthesis. However, application of the approach to photomorphogenesis, where the photoreceptors are present in much lower amounts, has met with limited success. For phytochrome, it did concentrate attention on the appropriate regions of the spectrum, but photochromicity (i.e. light-reversible changes in absorption properties) was the key to identification. With respect to the blue-light receptors, the chromophores are obviously some sort of yellow pigment, but it has proved impossible to distinguish between carotenoid and flavin candidates on the basis of action vs. absorption comparisons.

(2) saturation at relatively low fluence levels $(10–100 \text{ J m}^{-2})$;

(3) effectiveness of short exposure times (minutes).

Responses with these characteristics are termed **low-energy responses** (LER) and represent one particular (inductive) mode of phytochrome action.

Certain responses require much longer, or repeated, exposures to irradiation: the greening process itself requires several hours of irradiation, and light-induced leaf expansion is also brought about by continuous exposure to far-red radiation. This type of behaviour is attributed to a **high-irradiance reaction** (HIR), and represents another mode of operation of phytochrome. Responses initiated by the HIR are characterized by:

(1) requirement for high irradiances;

(2) requirement for long exposure times (hours) or repeated exposures;

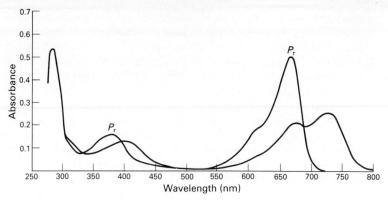

Figure 4.4 Absorption of purified rye phytochrome after saturating far-red (P_r) or red irradiation; absorption maximum P_r = 667 nm, P_{fr} = 730 nm.

(3) no reciprocity between time and fluence rate;

(4) no R–Fr reversibility;

(5) action maxima in one or more of the blue, red or far-red regions of the spectrum (see later).

The actual positions of the HIR action maxima were initially puzzling since they do not coincide with absorption by either P_r or P_{fr}. However, elegant experimentation by Hartmann in the 1960s demonstrated that phytochrome is indeed involved in the HIR. The inhibition of elongation in etiolated lettuce hypocotyls by continuous irradiation shows a far-red action maximum at λ 720 nm (Fig. 4.5a). Irradiation with λ 658 nm or with λ 768 nm alone has no effect on elongation; but when the radiation contains both these wavelengths together, then, at certain fluence rates, this composite radiation is as effective as λ 720 nm (Fig. 4.5b). The explanation derives from the fact that the absorption spectra of the two forms of phytochrome overlap, and thus different proportions of P_r and P_{fr} are present under different types of irradiation. Hartmann calculated that, at certain fluence rates, the synergistically effective λ 658 nm + λ 768 nm give rise to a ratio of active : total phytochrome (P_{fr} : P_{tot}) that is similar to that established by the maximally active radiation of λ 720 nm. Subsequent spectrophotometric measurements of phytochrome confirmed these calculations. Therefore, high-irradiance effects in the red to far-red region operate through the establishment of specific levels of biologically active phytochrome. There is still some controversy over the extents to which phytochrome and specific blue-light receptors contribute to high-irradiance responses in the blue region.

The HIR has been implicated in a wide range of responses, including seed germination, stem growth, leaf expansion and pigment synthesis. However, different species and different responses show what seem to

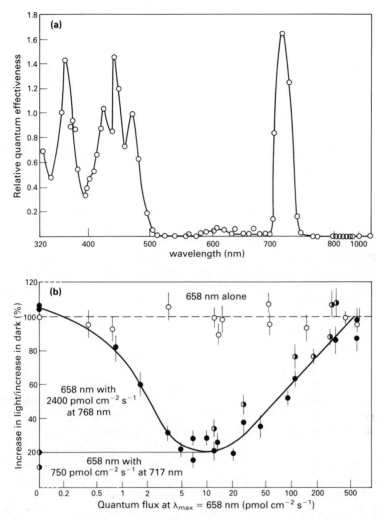

Figure 4.5 Light and hypocotyl growth in lettuce seedlings: (a) Action spectrum for inhibition of hypocotyl extension by continuous irradiation. (b) Effect of simultaneous irradiation with a fixed fluence rate at λ 768 nm and various fluence rates at λ 658 nm.

be different types of HIR. A tentative classification was proposed by Mancinelli (1980), based on the effectiveness of different spectral regions:

(1) *Action in three spectral regions* (B, R, Fr): this is characteristic of etiolated tissues – for example, continuous irradiation with these regions inhibits hypocotyl growth in most etiolated seedlings.

(2) *Action in two spectral regions* (B, R): this type of response is shown by green tissues – for example, responsiveness to continuous far-red irradiation is usually lost during de-etiolation.

(3) *Action in one spectral region*: this type of behaviour is shown in only a limited number of responses – for example, high irradiance blue light induces anthocyanin synthesis in sorghum, leaf-unrolling in rice and tendril-coiling in peas; high irradiance red inhibits hypocotyl elongation in green mustard seedlings (Fig. 6.7).

These differences in type of HIR may derive from differences in the extents to which phytochrome and blue-light receptors are involved. That is, some responses may only involve phytochrome (e.g. inhibition of hypocotyl elongation in green mustard seedlings); others may result from the action of only a blue-light receptor (e.g. leaf unrolling in rice seedlings); while the majority, like type (2) above, involve some form of interaction between phytochrome and blue-light receptors.

There is also the suspicion that the term HIR is used in different senses by different workers. At one extreme (Schopfer, 1984), the term is restricted to the effects of continuous far-red irradiation. In this sense, therefore, the HIR is considered to be something of a laboratory artefact, a feature of etiolated material with a high phytochrome content, and absent from green material – see the difference between (1) and (2) above. Many workers, however, seem to use the term HIR to refer to responses brought about by types of irradiation akin to natural light, such as high irradiance and long exposure.

As well as the LER and HIR, there may be a third mode of operation of phytochrome. Some American investigators (Mandoli and Briggs 1981) have recently found that growth reactions in oat mesocotyl and coleoptile are initiated by fluence rates of 10^{-4} to 3×10^{-2} µmoles m^{-2} s^{-1} – light levels four orders of magnitude lower than those which induce classical LER. These workers describe such behaviour as **very low fluence response**. These responses do not show R–Fr reversibility, and seem to be initiated when as little as 0.01% of phytochrome is in the P_{fr} form. Since irradiation by even the purest source of far-red available results in the production of some P_{fr}, the extreme sensitivity of the responses would account for the lack of reversibility.

4.2.3 The phytochrome molecule[3]

The phytochrome chromophore is a linear tetrapyrrole, which is linked to the apoprotein through a thioether bond from the vinyl group of

[3] See Pratt (1982, 1983), Quail *et al.* (1983), W. O. Smith (1983), Song (1984).

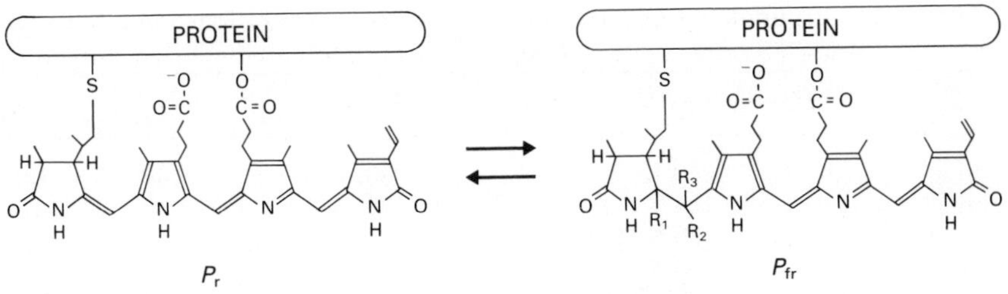

Figure 4.6 Molecular aspects of the phytochrome chromophore. The chromophore is linked to its apoprotein through a thioether bond from the vinyl group of ring A; it may also be linked through one of the propionic acid side-chains. The possible photo-transformation also depicted is based upon the addition of certain residues to the 4-5 double bond.

ring A (Fig. 4.6); one of the proprionic acid side-chains may also be covalently linked. The structure of P_{fr} is not yet clearly established, although the similarity of the shapes of the absorption spectra of P_r and P_{fr} suggest that the π electron system is conserved in both forms. P_{fr} may be a geometric isomer of P_r, or residues may be added to the methene bridge of ring A. In keeping with its rôle as a photoreceptor, the values for the absorption coefficients are relatively high:

$$\text{oat } P_r, \ \varepsilon_{664 \text{ nm}} \approx 7.6 \times 10^4 \text{ dm}^3 \text{ mol}^{-1} \text{ cm}^{-1};$$
$$\text{oat } P_{fr} \ \varepsilon_{724 \text{ nm}} \approx 4.6 \times 10^4 \text{ dm}^3 \text{ mol}^{-1} \text{ cm}^{-1}.$$

Values for the quantum yield of P_r phototransformation are also correspondingly higher than those for the P_{fr} photochange.

The apoprotein can constitute as much as 0.5% of the total extractable protein in etiolated oat seedlings, although the amounts are decreased around 100-fold in light-grown tissues. Phytochrome is thought to exist in solution as a dimeric protein, with one chromophore per monomer. Estimates of monomer size have undergone some change. Early studies indicated a molecular wieght of 120 K. However, the protein is highly sensitive to proteolysis, and readily loses a peptide fragment of MW 6–10 K. The MW of the intact native monomer is now recognized to be 124 K (i.e. the 120 K monomer is produced by proteolysis and, in fact, is a mixture of 114 K and 118 K units). The terms used to describe different preparations of phyto-chrome are shown in Table 4.2 (note the different absorption maxima for different forms of the molecule). Investigations of the apoprotein itself are continuing. It is a weakly acidic protein with an isoelectric point around pH 6.0. Hydrodynamically it behaves as an ellipsoid, and it has 20% of its structure in the form of α-helix, 30% as β-sheet and 50% of random coil. Such studies will hopefully throw light not only

Table 4.2 Terminology for different phytochrome preparations (adapted from Song 1984).

Term	MW	P_r λ_{max} (nm)	P_{Fr} λ_{max} (nm)
dimer	240 K	380, 666	400, 730
intact	124 K	378, 664	392, 719
large	120 K (114, 118 K)		
small	60 K		
fragment	20–42 K		
chromopeptide	2 K		

on how phytochrome is located within the cell, but also on functional aspects of the molecule.

The question fundamental to any consideration of phytochrome concerns the nature of the biologically active moiety. On the basis of circumstantial evidence, it is generally accepted that P_{fr} is somehow involved in this aspect. (For example, it is the P_r form which is present in etiolated seedlings; a brief flash of red light induces a biological response, which must be as a result of little change in the relative amounts of P_r but substantial change in the relative amounts of P_{fr}.) The actual nature of the light-induced changes in the phytochrome molecule are not yet clear. The protein configuration probably remains fairly similar, since there are no differences in sedimentation coefficients, electrophoretic properties or isoelectric points between the two forms of phytochrome. The molecule is more hydrophobic as P_{fr}. A recent model suggests that the isomeric change induced in the chromophore by red light results in its reorientation to expose a hydrophobic surface on the apoprotein (Fig. 4.7). Whether P_{fr} then combines with another component to produce a biologically active moiety, $P_{fr}.X$, is not known, but is required by many theoretical models of action. (Methods of phytochrome assay, involving properties either of the chromophore or of the apoprotein, were described in Box 4.2.)

4.2.4 Phytochrome Transformations[4]

Certain features of phytochrome behaviour thoroughly complicate any simple picture of its phototransformations and biological effectiveness.

The absorption spectra of the two forms of phytochrome overlap throughout much of their range (Fig. 4.4). This means that neither of the phototransformations can be driven wholly to completion, and under saturating irradiation, both forms are present in a **photoequilibrium**, designated by the symbol ϕ (phi). An earlier term is 'photostationary state'. While many authors use these terms inter-

[4] See Hayward (1984), Mohr (1984), H. Smith (1983a), Song (1984).

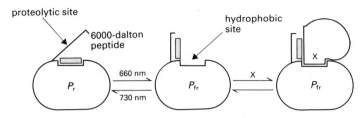

Figure 4.7 Model of the photo-transformation of phytochrome, shown in side view with respect to the chromophore crevice. 'X' represents an as yet unidentified cofactor.

changeably, the photoequilibrium and the photostationary state are nowadays considered to be not exactly analogous (see Glossary, and Hayward 1984).

The actual position of equilibrium is a function of wavelength and fluence rate. Thus, irradiation with different wavelengths at the same fluence rates establishes different photoequilibria: red light gives a maximum, but not total, conversion to P_{fr} ($\phi_{660} \approx 0.8$, i.e. 80% P_{fr}); far-red irradiation establishes an equilibrium in which there is still some P_{fr} present ($\phi_{730} \approx 0.01$, i.e. 1% P_{fr}); irradiation with other wavelengths also establishes characteristic equilibria (Fig. 4.8a). The value of ϕ is thus markedly influenced by the quantum flux ratio at λ 660 nm and λ 730 nm (Fig. 4.8b). (In fact, the largest effects of the R : Fr ratio on the photoequilibrium occur over the range of values found in the terrestrial environment, $\zeta = 0.1$–1.2, see also Table 3.3.) However, since the photoequilibrium is a function of fluence rate (N) as well as wavelength, the corollary is that the same photoequilibria can be achieved by different combinations of λ_{nm} and N. (For example, the photoequilibrium present in daylight is the same as that established by 452 μmole m^{-2}s^{-1} at λ 687 nm ($\phi = 0.61$); shadelight, $\phi = 0.27$, is equivalent to 17 μmole m^{-2}s^{-1} at λ 701 nm; recall, too, Hartmann's HIR experiments, where particular fluence rate combinations of λ 658 nm + λ 768 nm $\equiv \lambda$ 720 nm.) Therefore, *phytochrome can provide a measure of polychromatic radiation through effects of wavelength and fluence rate on the photoequilibrium; and the value of the photoequilibrium gives an indication of the level of biologically active phytochrome.* However, the relationship between the level of P_{fr} and the extent of biological response is not yet clearly established for all tissues and all responses.

Another complication to considerations of phytochrome behaviour in light is the occurrence of **intermediates** in the phototransformations of P_r and P_{fr}. These have been detected through the use of particular spectroscopic techniques such as *flash photolysis*, in which short pulses of light allow temporal distinction between different absorption forms, and *low temperature spectroscopy*, where distinction is achieved through differential effects of very low temperatures on reaction rates. All

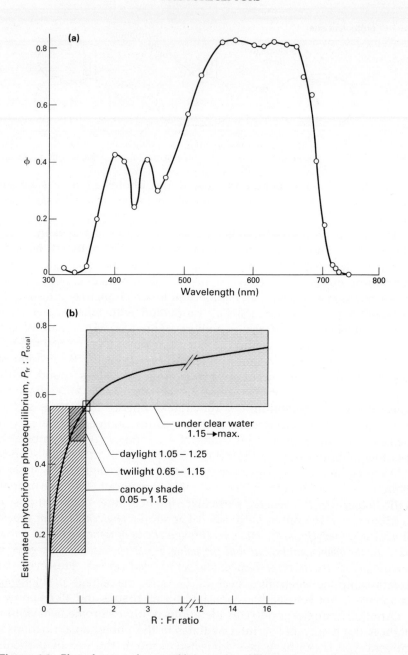

Figure 4.8 Phytochrome photoequilibria under different types of irradiation. (a) Photoequilibrium (φ) as a function of wavelength, determined from spectrophotometric measurements of etiolated mustard hypocotyls. (b) Photoequilibrium as a function of red : far-red ratio (ζ).

studies agree that (a) there are several intermediates; (b) photoconversions by red and by far-red radiation involve different, rather than reversible, pathways; and (c) some of the intermediates themselves are photoreactive. A scheme (Fig. 4.9) derived from low temperature spectroscopy studies illustrates the situation: photoexcitation of P_r is followed by dark relaxation through a number of intermediate forms to P_{fr}; photoexcited P_{fr} relaxes through a different set of intermediates to P_r. However, the actual details of the pathways have still to be elucidated – as does the significance of the intermediates to the biological activity of phytochrome. Since the thermal reactions must be much slower than the photoexcitations, then, *under continuous irradiation, it is likely that there is accumulation of these intermediate forms of phytochrome.* At the physiological level, an absorbance change at λ 543 nm in etiolated tissue exposed to simultaneous red and far-red irradiation has been equated with an accumulation of intermediates during phototransformation cycling.

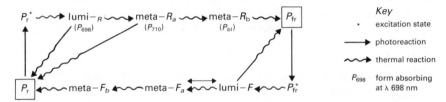

Figure 4.9 A scheme illustrating the involvement of intermediates in the phototransformations of phytochrome.

A third area of possible complication lies in the relatively slow **dark reactions** of phytochrome, particularly of the biologically active P_{fr}. (These physiological dark reactions should not be confused with the intermediate thermal reactions of the phototransformations.) *Dark destruction* describes the enzyme-catalyzed conversion of P_{fr} to a photoinactive form and involves degradation of the apoprotein. Rates of destruction vary between species: at 25°C, P_{fr} has a half-life of about 20 min in *Amaranthus caudatus*, compared to 50 min in *Helianthus annuus*. The reaction kinetics also vary: in dicots, phytochrome destruction shows first-order reaction kinetics (i.e. the rate is proportional to P_{fr} concentration); in monocots, the kinetics are zero-order (i.e. the rate is saturated even at low concentrations of P_{fr}). Therefore, particularly in dicots, the rate of dark destruction is indirectly influenced by light, through the level of P_{fr} established at photoequilibrium. The process may be of significance in the regulation of phytochrome activity. (Note, however, that these findings all come from investigations on etiolated plants which contain relatively large amounts of phytochrome.) *Dark reversion* is the spontaneous thermal reversion of P_{fr} to P_r. It occurs in some preparations of phytochrome *in vitro*, but its biological significance is very uncertain since, *in vivo*, it does not occur in monocotyledons or in the genera of the *Centrospermae*.

Recent findings indicate that the lower amounts of phytochrome found in green tissues are much more stable than phytochrome in etiolated material (see Ch. 6). That is, *the dark reactions of destruction and reversion occur to a much smaller extent in de-etiolated (Norflurazon-treated) seedlings*.

Thus, the simplest model of phytochrome behaviour may be formalized as shown in Fig. 4.10. Various attempts have been made, in terms of interactions between the processes shown in the figure, to account for the observed characteristics of phytochrome-regulated responses, in particular those of the HIR. One basic puzzle of the HIR is the irradiance dependency: increasing irradiance continues to increase the response, at fluence rates far higher than those required to saturate photoconversion. Several models have been proposed, and these all place different degrees of emphasis on different aspects of phytochrome behaviour. Briefly, the different models suggested are as follows:

(1) Biological activity by an excited form of P_{fr} (P_{fr}^*).
(2) Involvement of photo-cycling between the two forms of phytochrome. (This was invoked in the original Hartmann concept where the system was considered to 'count' the number of times the molecule cycled between P_r and P_{fr}; the response was related to the rate of interconversion and was thus dependent on irradiance at a critical photoequilibrium.)
(3) Competition between the 'P_{fr}-producing' light reactions and the 'P_{fr}-removing' dark reactions. (This accounts for the loss of HIR-Fr activity in green tissue where P_{fr} is much more stable.)
(4) Combination of P_{fr} with factor X to give the biologically active moiety, $P_{fr}X$.
(5) Biological activity from accumulation of a photo-intermediate.

It is intriguing that all these models (i.e. all explanations of phytochrome activity), imply, or even state, that P_{fr} itself is *not* the biologically active form of phytochrome!

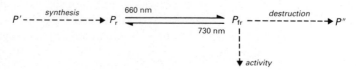

Figure 4.10 Basic model of phytochrome behaviour in terms of its involvement in reactions or processes.

Another related question concerns the actual significance of the photoequilibrium. According to some models, the instability of P_{fr} means that, in the longer term, the value of ϕ becomes independent of both wavelength and fluence rate; yet the extent of response is obviously influenced by both. Furthermore, it has been suggested that at low fluence rates (< 10 μmoles m^{-2}s^{-1}), the level of active phytochrome will be determined by the relative rates of synthesis and degradation, rather than by photoconversion. And at high fluence rates (> 300 μmoles m^{-2}s^{-1}), the significance of any equilibrium will be overshadowed by the accumulation of photo-intermediates.

4.3 BLUE-LIGHT RECEPTORS

4.3.1 Introduction[5]

There are a large number of responses to blue light (Table 4.3), whose action spectra suggest that they have a common type of photoreceptor (Fig. 4.11). Typically, they show an action maximum in the near ultraviolet (UVA), around λ 370–380 nm, and other maxima in the blue region, between λ 400–500 nm; individual responses differ in the relative effectiveness of the UV and blue regions.

For many responses, particularly in lower organisms, only this

Table 4.3 Some blue-light responses in plants.

Type of organism	Type of response
higher plants	inhibition of stem growth
	phototropism
	stomatal opening
	leaf unfolding (nyctinastic opening)
	leaf unrolling (rice)
	tendril coiling (pea)
	chloroplast development
	chloroplast orientation
	pigment synthesis
	enhanced protein synthesis
	enzyme synthesis
	enzyme activation (flavoenzymes)
lower plants	phototropism
	chloroplast orientation
	prothallus development
	conidiation
	sporulation
	pigment synthesis

[5] See Presti & Delbruck (1978), Briggs & Iino (1983), Senger (1982).

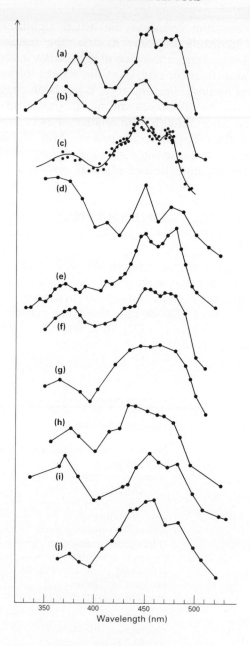

Figure 4.11 Action spectra for various blue light responses in different types of organism. Phototropism in (a) *Phycomyces*, (b) *Pilobolus* and (c) *Avena*; phototaxis in (d) *Euglena*; stimulation of carotenogenesis in (e) *Neurospora*, (f) *Fusarium* and (g) *Mycobacterium*; enhancement of respiration in (h) *Chlorella*; orientation of chloroplasts in (i) *Funaria*; circadian rhythm entrainment in (j) *Drosophila*.

blue–UV region is active in inducing the response (e.g. in processes of sporulation and prothallus formation, aspects of pigment synthesis and chloroplast development, and in phototropism). In these situations, it is obvious that blue light is operating through a non-phytochrome receptor. In many photo-responses of higher plants, however, both the blue and the red regions of the spectrum are effective. Since phytochrome is also activated by blue light, and since not all phytochrome-regulated responses are reversed by far-red irradiation, it is often difficult to judge whether such responses result from action of phytochrome or a blue-light receptor. (Indeed, for many years, the involvement of another receptor besides phytochrome was seriously doubted.) Various methods have been used to distinguish the action of a specific blue-light receptor, and can be illustrated by reference to case studies of hypocotyl growth inhibition under blue and red irradiation:[6]

(1) An ingenious approach involves the so-called *light-equivalent principle*. It will be remembered that specific phytochrome photoequilibria are established by irradiation with specific wavelengths, including those in the blue region. If a response is being induced solely through phytochrome, then irradiation treatments which give the same value of ϕ should give the same level of response. In Fig. 4.12, the percentage growth inhibition in response to radiation with different wavelengths is plotted against the phytochrome photoequilibria established by these wavelengths. Obviously, responses to the blue region (λ_s 400–500 nm) form a population that is distinct from those induced by far-red irradiation. Less formal variants of this 'light equivalent' method are also useful: when seedlings are grown under high irradiances of yellow sodium light, additions of small amounts of blue light have no effects on the value of ϕ, yet induce further inhibition of hypocotyl extension; again, differences between the fluence-response curves for blue and red light can provide a preliminary indication of different action mechanisms.

(2) There is often a *difference in the kinetics* of the responses to red and blue light. In many types of seedling, the inhibitory effects of blue light become apparent more rapidly than those of red light (Fig. 4.13).

(3) There can be a *spatial distinction* between regions of perception for blue and red light responses. The inhibitory effects of blue light on hypocotyl growth only occur if the hypocotyl itself is irradiated; red irradiation of the cotyledons inhibits growth of the hypocotyl.

(4) A *genetic distinction* can be made between the effects of blue and

[6] See Koorneef *et al.* (1980), Mohr (1984), Schäfer & Haupt (1983), Thomas (1981).

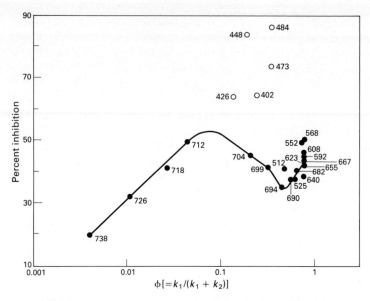

Figure 4.12 Distinction between effects of red and blue light on mustard hypocotyl growth by the 'light equivalent' method. The percent growth inhibition under different types of irradiation is plotted against the phytochrome photoequilibrium established by those wavelengths.

far-red irradiation; the growth of mutants which have lost their sensitivity to blue light is still inhibited by far-red irradiation.

4.3.2 Possible chromophores[7]

Comparisons of action spectra with absorption spectra implicate two types of pigment, carotenoids and flavins, as candidates for blue-light receptors. (Fig. 4.14). A *carotenoid* was first suggested, in 1936, to be the photoreceptor for phototropism. This type of pigment has a good match between absorption and action in the blue region – for example, β-carotene has three absorption shoulders at λ 480 nm, λ 450 nm and λ 420 nm; but, in aqueous solution, it has no UV absorption corresponding to the λ 370–380 nm action peak (Fig. 4.14*a*). Conversely, flavin has a clear absorption peak in the UV at λ 365 nm; but aqueous solutions of riboflavin have no trio of absorption peaks in the blue region (Fig. 4.14*b*). However, light absorption by a solid solution of riboflavin in ethanol at 77°K (Fig. 4.14*d*) does show an intriguing similarity to action spectra for responses in both the UV and blue regions. Equally, light absorption by a carotenoid in 54.7% ethanol shows a very close match to the action spectra of blue light mediated

[7] See Briggs & Iino (1983), Vierstra & Poff (1981a, 1981b).

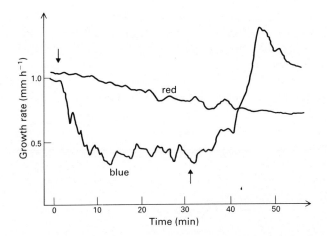

Figure 4.13 Distinction between the effects of red and blue light on mustard hypocotyl growth by kinetic analysis. Growth was measured continuously by a linear transducer; lights on and off at the arrows; red light = 8 W m^{-2}, blue light = 11 W m^{-2}.

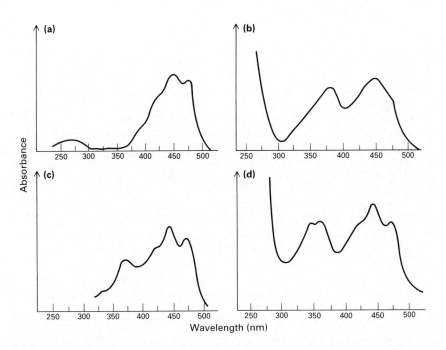

Figure 4.14 Absorption spectra of various preparations of carotenoids and flavins. (*a*) Aqueous solution of β-carotene. (*b*) Aqueous solution of riboflavin. (*c*) The carotenoid, lutein, in 54.7% ethanol. (*d*) Riboflavin in ethanol at 77°K.

75

response (Fig. 4.14c). (These last two examples may seem rather exotic situations, but then the state of aggregation and association of the receptor *in vivo* may also be complex.)

From considerations of the fine structure of the action spectra, some authors suggest that there is more than one type of blue-light receptor. Different systems are characterized by action spectra showing (see also Fig. 4.11):

(1) *One UV peak at λ 380 nm and three peaks in the blue region*, e.g. phototropism;

(2) *minor UV peak and two peaks in the blue region*, e.g. stimulation of carotenoid synthesis in *Neurospora* and of leaflet closure in the shade plant *Oxalis*;

(3) *one peak in the blue region*, e.g. enhancement of sporulation in *Penicillium*.

However, since the absorption characteristics of a chromophore can be markedly influenced by its molecular environment, these differences in action spectra may equally derive from differences in the particular states of organization of one general type of pigment.

Box 4.4   Other photoreceptors?

In many cyanobacteria and red algae, developmental responses to red and green light suggest the involvement of other photoreceptors besides phytochrome and receptors for blue light (Rudiger and Scheer 1983). The responses include effects on organism form – for example, in green light, *Fremyella diplosiphon* grows as filaments 460 μ long, with 40–60 cells per filament; under red light the organism has filaments of length 50 μ, comprising 10–15 cells. Another photoresponse of many marine dino-flagellates is that of *chromatic adaptation*, where the pigment complement of the organism is different in different spectral environments – a phenomenon that is particularly useful to organisms existing over a range of marine depths.

The putative photoreceptors which regulate these responses have received the general name *phycochromes*. The action spectra for the responses suggest the photoreceptor(s) to be some type of biliprotein – that is, a phycobilin chromophore tightly conjugated to a protein. Originally, and from analogy with phytochrome, it was thought that the responses regulated by the phycochromes showed photoreversibility by the green–red regions of the spectrum. A biliprotein showing photo-reversible changes in absorption has also been isolated. However, it may be that the photochromicity of the biliprotein is an artefact resulting from denaturation during isolation, and the opposite effects of red and green light on chromatic adaptation may simply represent antagonistic effects of separate photosystems.

Experimentation at the physiological level has also attempted to distinguish between carotenoid and flavin as receptor. A common approach is to 'remove' the type of molecule postulated to act as receptor in a response, and then see if light can still induce the response. For example, in lower organisms, mutants deficient in carotenoid synthesis can be produced; these mutants can have as little as 0.004% of the normal carotenoid concentration, yet still exhibit normal responses to blue light. Similarly, in higher plants, the herbicide Norflurazon has been used to inhibit carotenoid synthesis; again, seedlings treated with this chemical still display some photo-tropic response (Vierstra & Poff 1981b). On the other hand, chemicals like potassium iodide and phenylacetic acid are known to be strong quenchers of flavin photoexcitation; treatment of seedlings with either of these chemicals does decrease phototropic responsiveness (Vierstra & Poff 1981a). Therefore, present opinion tends to favour a flavin-type of molecule as a receptor for blue light.

(Particular flavoproteins can be directly influenced by light (Thomas 1981). For example, the flavoenzyme, nitrate reductase, is activated *in vitro* by blue light; this effect is enhanced by the addition of exogenous riboflavin, but not by β-carotene. However, whether such effects are relevant to actual physiological responses is not yet clear.)

4.3.3 Possible mechanisms of action[8]

General characteristics of blue-light mediated responses are:

(1) responsivity to a wide range of fluence rates, from less than 1.0 µmole $m^{-2}s^{-1}$ up to at least 400 µmole $m^{-2}s^{-1}$;
(2) relatively rapid responses (within seconds) to the presence and absence of light;
(3) effectiveness mainly on the actual tissue being irradiated.

Results from two separate areas of experimentation offer clues to possible mechanisms of action. Blue light induces stomatal opening through promoting K^+ uptake, and hence water uptake, into the guard cells. The driving force for the inward flux of K^+ is a light-induced outflow of protons; blue light induces a measurable efflux of H^+ from isolated protoplasts of guard cells. Therefore, does blue light have direct effect(s) on a membrane-located proton pump?

A different approach involves the spectrophotometric detection of absorbance changes induced by blue light in various tissues and tissue extracts. The actual changes involve increases in absorbance at λ 420–

---

[8] See Schmidt (1983), Senger & Briggs (1981), Thomas (1981), Zeiger (1983).

430 nm and at λ 550–560 nm, and may indicate photoreduction of a *b*-type cytochrome. They have been observed *in vivo* and *in vitro* in many fungal systems, and in a membrane fraction isolated from maize coleoptiles. The working hypothesis in these studies is that blue light induces the flavin-mediated photoreduction of a cytochrome on the cell membrane (with possible consequences for certain membrane properties?).

The possible existence of other photoreceptors besides phytochrome and blue-light receptors is considered in Box 4.4.

4.4 SUMMARY

(1) A pigment is an extended π orbital system; its capacity to absorb light is described by its absorption coefficient. Many compounds in plants absorb light, but only a few act in photosystems. The action spectrum of a photoresponse can indicate the nature of the photoreceptor.

(2) The existence of phytochrome was deduced from the R–Fr reversibility of physiological responses, and its presence was confirmed spectrophotometrically.

(3) Phytochrome consists of a photochromic chromophore conjugated to its apoprotein; photoconversion occurs through intermediates; saturating irradiation establishes a photoequilibrium between the two stable forms of phytochrome. Synthesis and destruction may also be involved in regulating phytochrome levels. In green tissues, phytochrome is present at much lower concentrations and seems much more stable.

(4) Phytochrome is capable of action in different operational modes:

LER: R–Fr reversibility, reciprocity, saturation at low fluence rates.

HIR: no reversibility, no reciprocity, irradiance and exposure period dependency.

Very-Low-Fluence: no reversibility, action at extremely low fluence rates.

(5) In many responses to blue light, activity of a separate receptor for blue light can be demonstrated in various ways: by the light-equivalent method, by kinetic analysis of the response and by genetic manipulation of the organism.

(6) Action spectra for blue light responses can be equated with absorption by flavins or by carotenoids; much physiological evidence favours a flavin-type photoreceptor. Blue light action may involve cytochrome reduction and changes in membrane permeability.

FURTHER READING

Cosens, D. & D. Vince-Prue (eds.) 1983. *The biology of photoreception*. SEB Symposia No. 36. Cambridge: Cambridge University Press.

Shropshire, W. & H. Mohr (eds.) 1983. *Photomorphogenesis*. Encycl. Plant Physiol. NS 16 A & B. Berlin: Springer.

Smith, H. & M. G. Holmes (eds.) 1984. *Techniques in photomorphogenesis*. London: Academic Press.

Wareing, P. F. & H. Smith (eds.) 1983. Photoreception by plants. *Phil. Trans. R. Soc.* **B303**, 345–536.

CHAPTER FIVE

Metabolic transduction and amplification of light signals

5.1 INTRODUCTORY COMMENTS

Any photomorphogenic response is made up of the processes of signal perception, signal transduction and amplification, and expression of morphogenic change. Transduction involves transformation of the environmental signal into a molecular or metabolic message, and may consist of a catenary sequence of events – a transduction chain during which there is substantial amplification of the message. For example, there can be a several thousand-fold increase between the number of quanta which elicit a response, and the final number of biological molecules 'altered' in the response; ·the apparent quantum yield of photomorphogenic change is considerably greater than one. Potential amplifying devices might well include regulators of metabolism and cellular activity, such as enzymes and plant growth regulators. Cell membranes, too, being both involved in the regulation of ion and metabolite transport and possible sites for photoreceptor localization, might be expected to have major rôles in the transduction and amplification of light signals.

The actual sequence of events between light signal and physiological response is not known for any transduction chain. The general principle underlying investigations in this area has been to assume that the shorter the time interval between light absorption and response, the greater the likelihood of a direct relationship between photoreceptor action and the response. Phytochrome action has been studied in two kinds of experimental situation: after brief inductive irradiation treatments, and under continuous irradiation. In the inductive situations, which involve R–Fr reversibility of a response, considerations of the time interval between perception and response are slightly more elaborate. If the period between red and far-red irradiation is extended beyond a critical point, the reversibility of the response is lost – the response escapes from photoreversibility. Such escape possibly represents completion of P_{fr} action; at that point, even if P_{fr} is phototransformed, the transduction chain is already initiated. The *escape times* for different responses vary from seconds (representing fast P_{fr} action) to several hours (representing delayed or continuous P_{fr} action). Further complexity derives from the fact that the response itself may take a long time to be expressed – that is, it may have a *latency period* of several hours or even weeks. Within this latency period though, the escape time is still measurable through assay of R–Fr reversibility. Thus, in this respect, there can be three types of phytochrome-mediated response:

(1) fast response (fast P_{fr} action, fast expression);
(2) slow response with a short escape time (fast P_{fr} action, slow expression);
(3) slow response with a long escape time (delayed or continuous P_{fr} action).

The speed of a response is an important consideration in each of the two major hypotheses concerned with the primary action of phytochrome:

(1) *action on cell membranes*, which could result in fairly rapid responses;
(2) *action on the genome*, that is, photomodulation of transcription or translation, which would probably show significant latency periods, though not necessarily long escape times.

5.2   PHYTOCHROME-MEDIATED RAPID RESPONSES[1]

There is a wide range of phytochrome-mediated rapid responses – that is, responses detectable within about five minutes of the onset of red irradiation and reversible by far-red. Many, though not all, of these responses do imply some change in membrane properties. However, it is difficult to establish the relevance of some of the rapid responses to physiologically significant processes.

The phenomenon of **pelletability** is one of the fastest responses to phytochrome action so far recorded. Brief red irradiation of etiolated seedlings results in a large increase in the proportion of phytochrome in the pellet fraction subsequently obtained by centrifugation of tissue homogenates. For example, in homogenates of dark-grown *Avena* seedlings, 90% of the phytochrome remains in the soluble, supernatant fraction; after red irradiation of intact seedlings for a few seconds,

[1] See Galston (1983), Haupt (1983), Quail (1983), Racusen & Galston (1983), Raven (1983), Roux (1983).

followed by immediate homogenization and centrifugation, 40–60% of the phytochrome is associated with the pellet. The effect is far-red reversible. It is still not certain that this behaviour represents a physiologically relevant association of active P_{fr} with a (membrane) receptor site.

A similar type of response, but detectable *in vivo*, is involved in the phenomenon of **sequestration**. Immunocytochemical assays show that

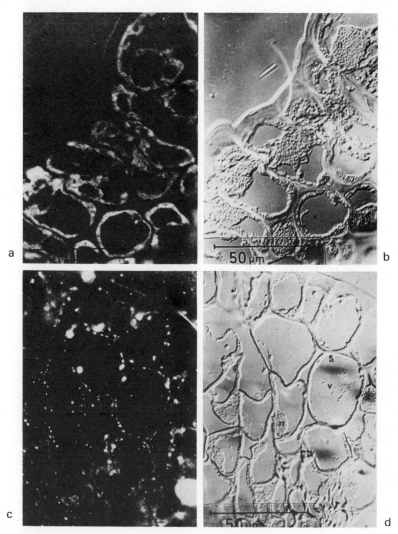

Figure 5.1 Sequestration of phytochrome. Phytochrome has been localized by immunofluorescence (white areas) in sections of *Avena* coleoptile cells: (a) dark-grown tissue; (c) after 3 min. irradiation with red light; (b) and (d) are the corresponding fields under Nomarski optics. a – amyloplast, n – nucleus, v – vacuole.

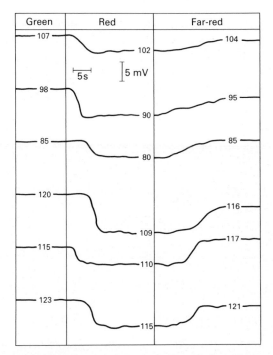

Figure 5.2 Changes in membrane potentials in parenchyma cells in six oat coleoptiles during sequential exposures to green, red and far-red irradiation. The numbers associated with sections on each trace are the mean absolute potentials in negative millivolts for that section of the trace.

in dark-grown tissue, phytochrome is diffusely distributed throughout the cytosol. After a few seconds of red irradiation, it is localized (sequestered) in discrete loci which are about 1.0 μm in diameter (Fig. 5.1). Again, since the loci are so far unidentified, it is not clear whether this response represents interaction with a membrane receptor, or is merely some form of self-aggregation.

Rapid, light-induced changes in **bioelectric potentials** are more obviously related to membrane-localized events. The first indication that phytochrome action produces electrical changes in plant tissues was the so-called *Tanada effect*. When etiolated roots are cultured in a (complex) nutrient solution, in a glass vessel whose walls are negatively charged with phosphate ions, they float freely while maintained in darkness. But after 30 seconds red irradiation, the root tips adhere to the glass; far-red irradiation brings about their release. Changes in the *surface potentials* of the roots are correlated with this adhesion and release. Light-induced changes in *transmembrane potentials* also occur in various tissues, although here the details of direction of change are in occasional conflict. In general, red irradiation causes a

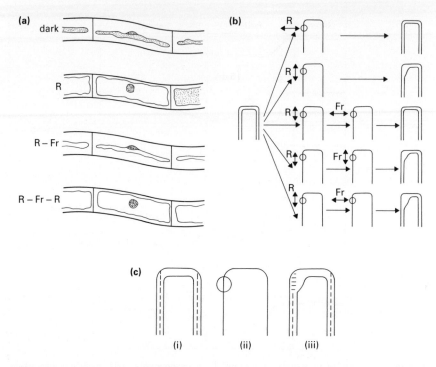

Figure 5.3 Chloroplast orientation in *Mougeotia* in response to red and far-red irradiation; the direction of irradiation is perpendicular to the page. (a) General responses to red (R) and far-red (Fr) irradiation. (b) Responses to polarized light (the electrical vector of the light is indicated by the double arrows). (c) Interpretation of the effects of red light on the orientation of phytochrome molecules.

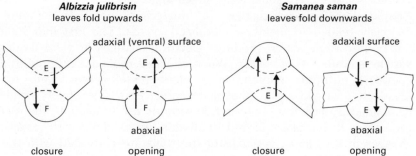

Figure 5.4 Summary of events during nyctinastic sleep movements in leaves of two species of legume (viewed through idealised transverse sections rachis and leaflets).

Box 5.1   Nyctinasty (sleep movement)

A nastic movement of an organ is induced by an external stimulus, but unlike a tropism, its direction derives from the anatomy of the organ, rather than the direction of the stimulus. The rhythmic sleep movements of leaves are termed *nyctinastic*. (Conversely, the light-regulated movements of the petals of many flowers, are generally referred to as photonastic.) Plants which show this behaviour have a bulbous zone, the *pulvinus*, at the base of the leaf petiole; there can be secondary pulvini at the bases of individual leaflets, and even tertiary pulvini in doubly compound leaves like those of *Mimosa*. Leaf movement is brought about by changes in the size of special cells within the pulvinus. These motor cells have numerous vacuoles and loose fibrillar walls: *extensor* cells take up K^+ ion and swell during leaf opening; *flexor* cells take up K^+ ion and swell during closure. The movement of K^+ into each cell type is driven by a H^+ efflux (cf. ion movements in stomatal guard cells, Section 4.3.3 and Box 6.3). Leaf opening and closure are thus both active events, and the apoplastic (i.e. membrane-crossing) ion fluxes associated with each cell type are independent. The actual events which occur during leaf movement are summarized in Figure 5.4.

Although nyctinastic movements occur in a very wide range of species, the underlying mechanisms have been most intensively studied in particular leguminous species (see Galston 1983). The movements are generally under the control of a *circadian rhythm*, and the actual effects of light vary according to the stage in the endogenous rhythm at which the light treatments are given (Ch. 8). Two photoreceptors are involved. Blue light, at the appropriate stage of rhythm, stimulates leaf opening. Phytochrome, however, has a more complex involvement which can be manifest in two ways. It is involved in the regular, daily entrainment of the circadian rhythm. It also affects the rates and extents of the actual leaf movements: red light (P_{fr}), at appropriate points in the rhythm, promotes closure through enhanced K^+ efflux from the extensors and influx to the flexors; treatment with far-red delays these processes. Thus, in the natural environment, leaf movements are driven by an endogenous rhythm, whose timing is regularly reset by phytochrome and whose actions on pulvini ion fluxes are reinforced by appropriately timed light treatments.

rapid depolarization within 5–30 s; repolarization under far-red irradiation takes longer (Fig. 5.2). Since such electrical changes must arise from electrochemical events related to ion movements, a case for their physiological relevance can be constructed: light brings about change in the cell membrane, which leads to a rapid redistribution of certain ions (e.g. H^+, K^+ or Ca^{2+}) with consequent effects on the activities of other biomolecules. That is, specific ions may act as *secondary messengers* in some phytochrome-mediated responses. In particular, it has been suggested that calcium may play such a rôle: P_{fr} may influence the concentration of cellular calcium and thus regulate the activity of a calmodulin-like protein, which in turn activates particular enzymes. (Red light does stimulate the uptake of Ca^{2+} into cells of the algae *Mougeotia*, and enhances the efflux of Ca^{2+} from protoplasts of *Avena*.)

Ion fluxes associated with **turgor changes** represent rapid responses to light in which the biological relevance is usually much clearer. The changes which occur during the nyctinastic sleep movements of leaves are of this type. The leaf movements derive from turgor-driven changes in volume of specialized motor cells at the leaf bases. The turgor changes themselves arise through redistributions of potassium ions. The overall regulation of these leaf movements is complex, and involves the interaction of phytochrome, a blue-light receptor and an endogenous rhythm (see Box 5.1). However, brief treatment with red light enhances the rate of leaf-folding, and involves increased rates of efflux of potassium from one type of motor cell (the extensors) and uptake of potassium by the other type (the flexors); far-red irradiation reverses these effects (and slows the rate of leaf-folding). Corresponding changes in the transmembrane potentials of these cells can often be measured even more rapidly, within seconds: treatment with red light leads to hyperpolarization of the flexor cells, while far-red irradiation gives depolarization.

The strongest evidence for the actual localization of phytochrome on a membrane is indirect, and derives from observations on the control of **chloroplast movement** in the filamentous green alga, *Mougeotia*. In this organism, each cell contains a single flat chloroplast whose orientation in low intensity light is mediated through phytochrome. (The orientation of chloroplasts in general, is more responsive to blue light.) In *Mougeotia*, after red irradiation, the flat face of the chloroplast is presented towards the light: far-red irradiation results in the edge being oriented towards the radiation source (Fig. 5.3a). In an elegant investigation which involved irradiation of different parts of the cell with microbeams of plane-polarized red and far-red light (Fig. 5.3b), Haupt showed that the phytochrome molecules involved in this response are:

(1) located outside the chloroplast itself;
(2) fixed in a non-streaming location (probably the plasmalemma);
(3) oriented in a dichroic pattern, with P_r parallel, and P_{fr} normal, to the cell surface (Fig. 5.3c).

The evidence for the last conclusion is that red light is most effective when its plane of polarization is parallel to the cell axis; far-red radiation is most effective when polarized in a plane at right angles to the cell axis (Fig. 5.3b). Thus, at least certain phytochrome molecules seem to be fixed in molecular arrays in or near the plasmalemma. However, the way in which phytochrome actually induces chloroplast movement is not yet clear. The dichroic orientations of the transition moments of the molecule result in a non-uniform distribution of P_{fr} around the cell after treatment with red light (see also Fig. 7.3a); the chloroplast 'moves away' from P_{fr} and thus tends to orient itself within this gradient of P_{fr}. The uptake of calcium into these cells is also stimulated by red light and the transduction chain for chloroplast movement may involve the interaction of calcium with the actomyosin machinery of cytoplasmic movement.

Other rapid effects of phytochrome are seen in the changed activities of certain enzymes, and these are described in the next section.

Rapid responses to light cannot easily be accounted for by action at the level of the gene. Borthwick and Hendricks originally proposed, in 1967, that phytochrome modified the functional properties of cell membranes, and clearly, many of the rapid responses do result from this. However, direct demonstration that phytochrome is physically located on a membrane, or that P_{fr} directly interacts with a membrane component, is still awaited.

5.3 LIGHT AND ENZYMES

5.3.1 Gene expression[2]

Control of development is ultimately concerned with the spatial and temporal regulation of gene expression. In the majority of light-regulated responses, the significant latency period before an effect becomes apparent is compatible with action at the level of the gene. The general protein complement of light-grown plant is significantly different from that of etiolated material, in both rate and pattern of protein synthesis (much of this difference is attributable to the transformation of etioplasts into chloroplasts).

The original suggestions that light may regulate gene expression (made by Hans Mohr in the mid-1960s) assigned the point of possible photocontrol to the level of transcription. However, little evidence was found initially for light-induced changes in transcription. This may have been due to technical difficulties associated with the requirement for quantitative isolation of different mRNAs. The presence of poly-A sequences on mRNA now allows this to be accomplished by affinity

[2] See Kendrick (1983), Lamb & Lawton (1983), Tobin & Silverthorne (1985), Schopfer (1984).

chromatography, and recent results indicate that transcription can indeed be affected by light. For example, production of chlorophyll a-apoprotein (Ch. 6) seems to be photoregulated at this level. This protein is located in the chloroplast but is encoded in the nuclear genome and synthesized on cytoplasmic ribosomes. Increased activity for apoprotein-mRNA appears in the light. The small subunit of the CO_2-fixation enzyme, ribulosebisphosphate carboxylase, is another chloroplast protein which is synthesized in the cytoplasm and for which increased mRNA activity appears after a light treatment. (In fact, in this case, although there is a significant lag period before the appearance of the increased mRNA activity, the response has one of the shortest escape times for R–Fr reversibility.) Another example of photocontrol of transcription is seen in the increased levels of rRNA in light-treated material.

A considerable body of evidence also exists for the photoregulation of certain processes at the level of translation. The proportion of ribosomes in the polysome fraction is increased by red light. This effect, which is reversed by far-red radiation, is only partly accounted for by increased mRNA activity. Photocontrol may also be exerted through a light-induced selective recruitment of existing mRNA into the polysome fraction.

Light probably also affects post-translational aspects of protein synhesis. Processes of leader sequence excision, glycosylation, prosthetic group attachment, oligomer assembly and inactivator degradation are all areas of potential photoregulation.

Thus, light is clearly involved in the regulation of gene expression at a number of levels. Many different molecular control mechanisms operate, probably even in the expression of any one gene. However, there is, as yet, no evidence for the direct interaction of phytochrome, or any other photoreceptor, with the genetic material itself.

5.3.2   Photoregulation of enzymes[3]

Examples of photoregulated enzymes can be found over a wide range of metabolic activities, including photosynthesis, photorespiration, pigment synthesis, fat metabolism, starch degradation and nitrate reduction. The effects of light on an enzyme are often described as *photomodulatory*, since light (merely) brings about changes in an existing enzyme or influences existing rates of synthesis or degradation. That is, unlike enzyme induction in bacteria, there is no case known in higher plants where light switches on the synthesis of a 'light-specific'

[3] See Buchanan (1980), Holmgren (1985), Newbury (1983), Schopfer (1977), Thomas (1981).

enzyme. Furthermore, even when material is transferred to conditions of continuous irradiation, light-modulated changes in enzyme activity are generally transient, indicating that other endogenous regulatory mechanisms are also involved.

The lag periods for the appearance of light-induced effects on different enzymes vary greatly, from minutes to several hours, suggesting that different mechanisms are involved in their photo-regulation. A change in the level of enzyme action can arise in two ways:

(1) change in *amount* of enzyme, through effects on enzyme synthesis or degradation;
(2) change in *activity* of enzyme by processes of activation or inactivation.

In practical terms, it is rarely straightforward to distinguish between these two situations.

Investigations into possible light-induced change in the rate of synthesis of an enzyme typically proceed through various levels of evidence as follows: ('rate of synthesis' strictly should read 'relative rates of synthesis and degradation', since it is also not easy to distinguish between an increased rate of synthesis and a decreased rate of degradation):

(1) Use of *inhibitors* of protein synthesis (e.g. cycloheximide) or RNA synthesis (e.g. actinomycin D, cordycepin) can only provide preliminary indications, since even positive effects are subject to serious difficulties of interpretation. (For example, if such an inhibitor does prevent light-induced change in an enzyme, it means that protein synthesis is a prerequisite for the change, but not necessarily synthesis of that particular enzyme; another regulatory protein may be involved.)
(2) Measurement of the *in vivo* incorporation of labelled amino acids into newly-formed protein can indicate the possible extent of synthesis of a particular enzyme. *Density-labelling* methods use amino acids labelled with a stable heavy isotope, usually ^{18}oxygen or deuterium. (This is often done simply by incubating the tissue or seedlings in 'heavy water', thus labelling amino acids as they are hydrolysed from storage protein.) The extent of 'heavy amino acid' incorporation into the enzyme is determined from the distribution of enzyme activity in a centrifuged density gradient. *Radiolabelling* involves incubation with a radioactive amino acid, often ^{35}S-methionine, followed by specific immuno-precipitation of the enzyme and determination of its radioactivity.

Box 5.2 Anthocyanin synthesis

It is a common observation that apples are redder on the sunnier side of the tree. Indeed, increase in the rate of anthocyanin synthesis was one of the earliest biochemical observations of the effects of light on plant development (Hahlbrock & Grisebach 1979, Mancinelli 1983).

Anthocyanins are flavonoid derivatives of cinnamic acid, which is itself formed from the amino acid, phenylalanine. The metabolic pathway leading to flavonoids begins with three reactions common to the synthesis of all major phenylpropanoids. This central sequence of three reactions is therefore termed *general phenylpropanoid metabolism*, and is also the starting sequence for lignin and coumarin synthesis:

(1) cinnamic (2) hydroxycinnamic (3) 4-coumarate.CoA

$$\swarrow \quad \downarrow \quad \searrow$$

phenylalanine $\longrightarrow$ acid $\longrightarrow$ acid $\longrightarrow$ lignins flavonoids coumarins

Enzyme (1) catalyses the initial deamination of phenylalanine, and is called *phenylalanine ammonia lyase* (PAL).

Besides its obvious commercial importance, anthocyanin synthesis also acts as a model system for the biochemical investigation of photomorphogenesis (the end-product itself can be extracted simply and assayed spectrophotometrically). In general, the effects of light on anthocyanin production are characteristic of the HIR: full expression of the response requires prolonged exposure to relatively high fluence rates. In most systems, two photoreceptors seem to be involved: an exposure to blue light increases the effectiveness of P_{fr}. In sorghum, pretreatment with blue light is obligatory before an effect of phytochrome can be seen. The mustard seedling is unusual in expressing maximum synthesis solely in response to phytochrome action. Studies on the photoregulation of the enzymes involved in flavonoid synthesis have been restricted to those of the initial general propanoid sequence; PAL has been the subject of particularly intensive investigation (see text).

Anthocyanins occur as water-soluble vacuolar pigments in all orders of higher plants except the Centrospermae (the red pigment of beetroot is a nitrogenous betacyanin). The capacity for anthocyanin synthesis is a function of genetic and environmental factors: the latter include nutrition, water supply, wound effects and infection, as well as light and temperature. Age is also a factor; greatest production is seen in young seedlings or mature flowers and fruits. The general restriction of anthocyanins to epidermal or hypodermal cells suggests that they function as protectants against UV damage, and as visual attractants to aid pollination and seed dispersal by animals.

In both methods, it is comparison of the extent of incorporation in the presence and absence of light that is important (i.e. light does not switch on the synthesis of new enzyme, it only modulates the amount of enzyme).

(3) *Immunoassay* techniques can be used to determine the specific levels of individual enzymes.

(4) Changes in protein synthesis can often be reflected in changes in levels of *specific mRNAs*. Relative changes in mRNA activity are detected by analyses of the products after translation of the total mRNA fraction in an *in vitro* system. Alternatively, specific mRNAs can be investigated by the use of 'hybridization probes' (i.e. a cloned DNA sequence complementary to the mRNA).

Future work in this area will depend very heavily on such immunological and recombinant DNA techniques.

The other means of enzyme regulation involves processes of enzyme (in)activation. These effects can be apparent much more rapidly, often within minutes, but it is arguably even more difficult to demonstrate that an effect is unequivocally due to enzyme activation – there is such a large variety of possible mechanisms, and experimental protocols and 'rules of evidence' have not really been formalized yet. Mechanisms include changes in enzyme configuration due to the types of

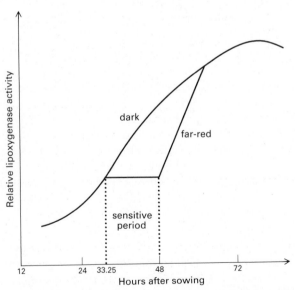

Figure 5.5 Time-course of the changes in lipoxygenase activity in mustard cotyledons in darkness or under continuous far-red irradiation.

post-translational modifications mentioned previously, or to light-mediated interactions of the enzyme with another regulatory moiety. (The background roles of some light-regulated enzymes which have been the subject of particular study are described in Box 5.2.)

5.3.3 Phytochrome and enzymes[4]

The activities of a large number of enzymes are affected *in vivo* by red irradiation. Even in those cases where the operational criteria clearly indicate phytochrome action, differences in other aspects of the responses suggest that quite different types of regulatory mechanism are involved. The behaviour of three illustrative examples is described.

Nitrate reductase (NR) is a key enzyme in the nitrogen supply route of most plants, and its synthesis is induced by the substrate (NO_3^-). The activity of the enzyme is also strongly affected by light. Phytochrome involvement has been demonstrated in a number of systems, including pea buds, cereal leaves and in mustard cotyledons where NR activity is stimulated within five minutes of the onset of red irradiation. Enzyme activation is thought to play a major role in these effects, probably through the mediation of other regulatory proteins. (In some cases, there may also be increased synthesis of NR as a result of phytochrome action. And other photoreceptors can also be involved in regulation of NR activity – for example, in peanut, blue light inhibits the action of an inactivator of NR, and thus enhances NR activity.)

Lipoxygenase (LOG) catalyses the oxidation of certain types of unsaturated fatty acid. Its photoregulation shows several interesting features. Expression of LOG activity is *prevented* by phytochrome action. However, in certain systems there is a limited period of sensitivity to the action of light. For example, in mustard cotyledons, LOG activity increases during growth in darkness (Fig. 5.5). Exposure to continuous far-red irradiation has no effect on this increase until 33 hours (at 25°C) from the start of imbibition; from 33 to 48 hours, continuous far-red irradiation prevents further increase in activity, and then suddenly ceases to have any effect. There is some controversy as to whether, in this case, light prevents further synthesis of LOG (phytochrome would here be repressing gene expression), or whether light simply blocks some final activation step in enzyme production. A further feature of phytochrome action in this system is that the relationship between response and level of P_{fr} suggests the operation of some threshold mechanism; if the level of P_{fr} exceeds 1.25% of P_{tot}, the effect on LOG activity is fully expressed; if P_{fr} falls below 1.25%, the effect is completely lost. This intriguing relationship perhaps indicates the

[4] See Schopfer (1977, 1984).

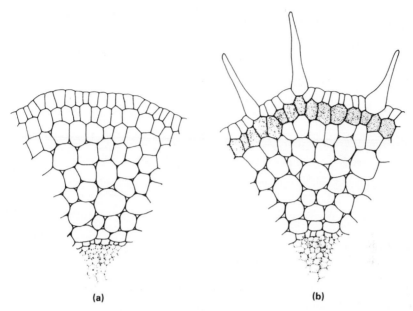

(a)   (b)

Figure 5.6 Spatial sensitivity of response to phytochrome. Transverse sections of mustard hypocotyls grown in darkness or under far-red irradiation. Anthocyanin production is confined to the hypodermal layer, and only certain epidermal cells develop into hair cells.

requirement for a particular number of active phytochrome molecules in each cell.

Phenylalanine ammonia lyase (PAL) activity is regulated by more than one photoreceptor and by more than one type of mechanism. Phytochrome is clearly involved: PAL activity is increased both by short irradiation periods with typical R–Fr reversibility and by continuous irradiation, with HIR action maxima characteristic for the particular system. In some cases, phytochrome seems to affect enzyme synthesis. For example, in mustard cotyledons, the light-induced increase in activity is inhibited by cycloheximide and actinomycin D; there is greater incorporation of density-labelled amino acids into the enzyme in light than in darkness; and there are large increases in mRNA activity for PAL In light. However, regulation of activity in the completed enzyme also seems to be accomplished through some kind of additional phytochrome-mediated mechanism which involves enzyme inactivation. (And in gherkin hypocotyls, there is a blue light-mediated activation of PAL activity.)

Thus, phytochrome seems to be involved in the photoregulation of enzymes through a variety of mechanisms. Overlaid on this, is another level of complexity. It has already been suggested that the transient nature of the responses of many enzymes, even to continuous

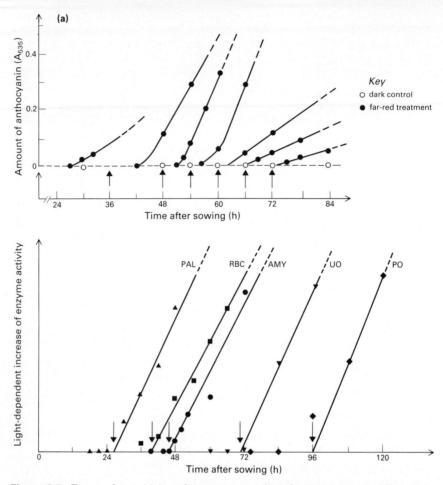

Figure 5.7 Temporal sensitivity of response to phytochrome, seen as differential development of enzyme activities under continuous irradiation. (a) Time-course of responsiveness of anthocyanin synthesis in mustard; dark-grown seedlings were transferred to continuous far-red irradiation at the times indicated by the arrows; O – dark control, ● – far red. (b) Temporal pattern of competence for phytochrome induction of enzyme activities in cotyledons; mustard seedlings were irradiated from sowing with continuous red or far-red light, and the 'starting points' for the appearance of the different activities are arrowed.

irradiation, may be due to the operation of other endogenous regulatory mechanisms. The variation in sensitivity of LOG to phytochrome action provides a clear example of this. Such behaviour is characteristic of many of the interactions between phytochrome and enzymes. *Spatial sensitivity* to the phytochrome-mediated action of light is seen in the restriction of anthocyanin synthesis to the hypodermal layer in mustard hypocotyls (Fig. 5.6). Other examples of *temporal sensitivity* in the responses of enzymes towards phytochrome are seen in the changing responsiveness of anthocyanin synthesis to far-red irradiation (Fig. 5.7a), and in the sequential development of different enzyme activities in mustard cotyledons under continuous irradiation (Fig. 5.7b). That is, in its actions on enzyme development, phyto-chrome seems to trigger responses which are already spatially and temporally preprogrammed by some endogenous regulatory system.

5.3.4 *Other effects of light on enzymes*[5]

As indicated in the previous section, NR and PAL can be photoregulated specifically by blue light. Some of these effects may be mediated through the actions of other regulatory molecules on the activity of the enzyme. However, it is also conceivable that blue light could directly affect the activity of certain enzymes. *Flavoenzymes* carry a flavin molecule as coenzyme or prosthetic group. Obviously, the absorption spectrum of a flavoenzyme is closely similar to action spectra for blue-light responses and, indeed, spinach NR, which is itself a flavoenzyme, can be activated *in vitro* by blue light. However, the biological significance of such effects remains to be established.

Light is capable of stimulating the activity of many enzymes *in vivo* by a mechanism which does not involve either phytochrome or a blue-light receptor. It is another type of so-called protein-mediated control, where light brings about a change in some regulatory protein, which in turn affects the activity of an enzyme. Examples of this method of regulation are seen in the rapid changes in activity of some of the photosynthetic enzymes in response to light and darkness. The effects of certain chemicals on these responses indicate the basis of the mechanism: an inhibitor of photosynthetic electron transport, DCMU (dichloromethyldiphenylurea) prevents the light-stimulation of activity; a thiol reducing agent, DTT (dithiothreitol), mimics the effects of light and removes the DCMU-inhibition. That is, light acts through the photosynthetic pigments and their electron transport system to reduce thiol groups on an enzyme. Reduction of these sulphur groups

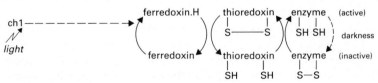

Figure 5.8 Diagrammatic representation of the action of thioredoxin in regulating enzyme activity.

[5] See Buchanan (1980), Holmgren (1985), Thomas (1981).

results in the protein adopting a configuration in which it is enzymatically active. A small sulphur protein, thioredoxin, is the connecting redox hydrogen-carrier in the system (see Fig. 5.8). This mechanism operates on at least four enzymes of the Calvin cycle: fructose-di-phosphatase, sedoheptulose-di-phosphatase, phosphoribulokinase and glyceraldehyde-phosphate dehydrogenase. It also seems to be involved in the photoregulation of activity in some non-photosynthetic enzymes, including PAL. It may be that there are specific forms of thioredoxin for different enzymes.

5.4 LIGHT AND GROWTH REGULATORS[6]

5.4.1 Introduction

So-called plant hormones occupy a central rôle in the control of growth and development. They are active in minute amounts, yet cause large changes in rates and patterns of growth. Therefore, any mechanism through which an environmental signal could induce a change in hormonal activity would be a means of achieving powerful amplification in a growth response. The natural plant growth regulators include auxin, more than 60 native gibberellins (although not all may express hormonal activity), several endogenous cytokinins, abscisic acid, ethylene and various inhibitors with regulatory rôles.

The effectiveness of a growth regulator can be modified in various ways:

(1) Change in *amount* of substance, through effects on its rates of synthesis–degradation. (For example, IAA can be degraded by a peroxide-based IAA-oxidase. Isoenzymes of this peroxidase are readily inducible, and their types and numbers differ according to the physiological status of the tissue.)

(2) Change in *activity* of substance, through processes of activation (release) or inactivation (binding). (For example, supraoptimal IAA levels can be adjusted by sequestering the auxin in the form of an inactive complex.)

(3) Change in *sensitivity* of the tissue to the hormone. That is, hormones only express activity when specific cells (target cells) are physiologically primed to respond. Variation in sensitivity could derive from differences in number of cellular receptor sites or from differences in affinity of receptors for the hormone.

[6] See Black & Vlitos (1972), DeGreef & Fredericq (1983).

5.4.2   *Light and gibberellins*[7]

Amount of gibberellin
Many physiological responses to light can be duplicated by the exogenous application of gibberellin. For example, application of gibberellin can induce:

(a)   dark germination in (some) light-sensitive seeds (e.g. lettuce);
(b)   flowering in (some) photoperiodically sensitive plants, particularly long-day species;
(c)   leaf expansion in (many) dark-grown seedlings.

Such effects suggest that one action of light may be to increase the levels of endogenous gibberellin, a hypothesis that can be directly tested by measuring the levels of gibberellin after a light treatment. One well-characterized system utilizes the developing cereal leaf. Light causes the tightly rolled primary leaves to unfold, through enhanced cell expansion in the adaxial mesophyll. The response shows *R–Fr* reversibility, and can also be induced by gibberellin alone. In etiolated barley and wheat seedlings, there are transient increases in endogenous gibberellin after various types of red light treatment. After a 30-minute irradiation period, there is an increase in gibberellin 20 minutes later; this may be the result of increased synthesis of the hormone, since it does not occur in tissues treated with inhibitors of gibberellin biosynthesis such as AMO 1618 (4-hydroxy-5-isopropyl 2-methylphenyl trimethyl ammonium chloride 1-piperidine carboxylate) or CCC (2-chloroethyl-trimethyl ammonium chloride). After a five-minute treatment with red light, increased gibberellin levels can be detected within five minutes; the rapidity of this response, and the ineffectiveness of it on AMO 1618 or CCC, suggest that it represents release of hormone from bound or sequestered forms. In this respect, plastids are considered to be an important intracellular source of gibberellin, and it is suggestive that red light also induces release of gibberellin from isolated etioplasts *in vitro*. Therefore, a phytochrome-mediated efflux of gibberellin from plastids, followed by a further light-induced synthesis of gibberellin, may be causally involved in cereal leaf-unrolling. (However, other regulatory processes must also be involved; application of cytokinin similarly induces unrolling in darkness; and the actual unrolling response can still be affected by far-red irradiation two hours after the inductive red light treatment, i.e. long after the transient changes in gibberellin levels.) Increases in endogenous gibberellin levels after light treatment also occur in other systems; for example,

[7] See Black & Vlitos (1972), DeGreef & Fredericq (1983).

imbibing lettuce seed show increased levels 60 minutes after red irradiation; and increased levels of gibberellins, or of rates of gibberellin turnover, are common responses to photoperiodically inductive long days in many species.

On the other hand, there is considerable evidence against the view that light stimulates gibberellin production. Many light-induced responses are neither duplicated nor even affected by exogenously-applied gibberellin; the development of amylase activity in mustard cotyledons is strongly stimulated by red light, but is absolutely unaffected by exogenous gibberellin; the dark germination of the 'long-day sensitive' seeds of *Kalanchoe* and *Begonia* cannot be brought about by gibberellin. Further, many light responses are actually reversed by gibberellin. The light inhibition of stem extension, particularly evident in dwarf varieties of peas, maize and rice, can be negated by application of gibberellin, as can the light-induced unfolding of the plumular hook in beans and the light stimulation of flavonoid synthesis in peas.

Gibberellin and tissue sensitivity

The classic effects of gibberellin in relieving the light inhibition of internode extension in peas originally prompted suggestions that light may act in this situation by lowering the levels of endogenous gibberellin. However, there are no significant differences in the amounts of the two major gibberellin fractions (GA_1 and GA_5) in dark-grown or light-treated pea plants of either the dwarf or tall varieties. Instead, red irradiation results in a decreased ability of the tissue both to respond to exogenous GA_5 and to accumulate radio-labelled GA_5. These effects are thought to be due to a decreased capacity of the tissues themselves to bind the hormone to specific gibberellin receptors; the three-hour lag period before the effect of light on hormone accumulation becomes apparent suggests that light reduces the number of receptor sites, rather than their affinity for the hormone. Thus, another action of light could be to lower the physiological responsiveness of the tissue to endogenous gibberellin.

This concept of growth pattern being determined by variation in the ability of a tissue to respond to modulator molecules may be the basis for many phenomena in (photo)morphogenesis.

5.4.3 Light and other hormones[8]

Auxin

There are interactions between light and many auxin-regulated responses – for example, auxin-stimulated elongation of stem sections

[8] See Black & Vlitos (1972), DeGreef & Fredericq (1983).

is sensitive to red and far-red irradiation, auxin-induction of lateral root formation is inhibited by red light, and tropic responses are strongly influenced by pretreatment of the organ with red light. Light treatments often result in decreased levels of extractable auxin in tissues (e.g. in pea and bean shoots). However, exogenous application of IAA rarely reverses a physiological response to light, even in situations that are associated with a measured decrease in endogenous auxin. Furthermore, in many systems light treatment also results in decreased IAA-oxidase activity, a situation that could be expected to result in an increased, rather than decreased, level of auxin. And the auxin content of oat mesocotyl increases in red light, although growth is strongly inhibited. Therefore, except perhaps in the special case of phototropic responses (Ch. 7), the general relationship between light and auxin is not clear.

Ethylene
In many cases, light and ethylene induce opposite growth responses. Application of ethylene negates the stimulatory effects of light on the rate of leaf expansion in de-etiolating pea seedlings. And light treatments of etiolated tissues are often followed by decreased rates of endogenous ethylene output. This latter type of effect has been particularly studied in relation to the behaviour of the bean plumular hook: ethylene promotes hook-closure, light induces opening, and ethylene production is lowered in light. It has been suggested that light inhibits ethylene synthesis. However, in anthocyanin production, light and ethylene act in the same (stimulatory) direction. In green *Marchantia* thalli, perturbation of ethylene output can also be achieved through effects on respiratory and photosynthetic metabolism, and it has been concluded that, at least in that system, changes in ethylene output are a by-product, rather than a mediator, of phytochrome-induced changes in metabolism.

Cytokinin
In a range of superficially similar responses, such as leaf expansion in a number of species, leaf unrolling in cereals and frond multiplication in *Lemna*, application of kinetin can substitute for treatment with red light: but in the germination of light-sensitive seeds, kinetin serves only to increase the sensitivity of the process to lower light dosages. In several systems, such as during imbibition of lettuce seed and in frond multiplication of *Lemna*, red light treatments are followed by changes in levels of endogenous cytokinin. Although all these processes obviously involve steps that are regulated by cytokinin, quantitative investigation of the relationship between light and exogenous kinetin generally reveals either an additive or synergistic type of interaction; both of

these situations indicate that the two factors probably have independent actions at different points within a process.

Inhibitors

Levels of *abscisic acid* are greater in dwarf varieties of rice than in tall, and it was originally thought that this inhibitor might also be involved in the light-regulation of stem extension. However, there are no significant differences in the levels of abscisic acid in light-treated and dark-grown pea shoots. Photoperiodically induced bud dormancy generally involves changes in inhibitors, including abscisic acid, with short days giving increased levels. *Xanthoxin*, chemically derived from xanthophyll and related to abscisic acid, has strong inhibitory effects on growth, and is present in greater concentrations in light-treated tissues. However, its status as a natural growth regulator is controversial. Similarly, other inhibitory compounds, such as flavonoids and phenols, though also present at increased concentrations in light, may not have rôles in natural growth regulation.

Concluding comments

Despite the considerable research effort in this area, the general relationship between light and any hormone is not clear. Light and hormones do interact to control growth. Some effects of light can be duplicated by application of hormone, particularly gibberellins; in other systems, the effects of light and hormones seem to be antagonistic. Significant changes in amounts of hormones often occur after light treatments. However, there is, as yet, no unequivocal evidence for a direct involvement of any plant growth regulator in the causal sequence between photoreceptor action and photomorphogenic response. In many responses, light may alter, positively or negatively, the *responsiveness* of cells to hormone, rather than have direct effects on hormone metabolism.

5.5 SUMMARY

(1) A light signal is transduced into a metabolic message and amplified to give a photomorphogenic response; the shorter the time interval between light absorption and the response, the closer may be events to photoreceptor action.

(2) Many rapid responses to phytochrome action involve changes in the functional properties of the cell membrane (changes in membrane potential, ion flux and cell turgor). There are indications that phytochrome may interact with membrane components (pelletability, sequestering), and is actually located on the

membrane (control of chloroplast movement in *Mougeotia*).

(3) Many of the slower responses to phytochrome seem to be mediated through selective gene expression, at the levels of both transcription and translation.

(4) Enzymes and growth regulators act as amplification mechanisms in the transformation of a light signal into a photomorphogenic response.

(5) The level of enzyme action can be regulated by light in various ways; phytochrome can influence enzyme synthesis and activation; light may act directly on flavoenzymes; the photosynthetic light-processing machinery can also exert an influence on enzyme activity, through thioredoxin.

(6) Light and growth regulators interact in many developmental responses. Levels of hormone may be regulated by effects on synthesis and activation. The sensitivity of the tissue to the hormone may also be regulated by light.

FURTHER READING

DeGreef, J. A. & H. Fredericq 1983. Photomorphogenesis and hormones. In *Photomorphogenesis*, Encycl. Plant Physiol. NS 16A, W. Shropshire & H. Morh (eds.), 401–27. Berlin: Springer.

Quail, P. H. 1983. Rapid action of phytochrome in photomorphogenesis. In *Photomorphogenesis*, Encycl. Plant Physiol. NS 16A, W. Shropshire & H. Mohr (eds.), 178–212. Berlin: Springer.

Schopfer, P. 1984. Photomorphogenesis. In *Advanced plant physiology*, M. B. Wilkins (ed.), 380–407. London: Pitman.

CHAPTER SIX

Growth and development: Photomorphogenesis

6.1 INTRODUCTORY COMMENTS[1]

Photomorphogenesis is an essential feature of all stages in the ontogeny of most plants – probably no higher plant can complete its life cycle without developmental control by light. The delineation of 'photomorphogenesis' is not quite straightforward, since although it describes the normal development of a green plant in light, the term also embraces other regulatory aspects of light on plant form and orientation, including such non-developmental processes as photo-tropism, photonasty, and so forth. That is, some photomorphogenic responses are irreversible and may involve gene expression, such as leaf expansion and pigment synthesis; others are reversible and do not involve action at the gene level, such as chloroplast orientation and enzyme activation. Phytochrome acts in both types of response.

Photomorphogenesis includes numerous individual responses which are highly specific for a particular organ at a particular time. Examples of spatial and temporal specificity in responses to light have already been seen in enzyme and hormone activities (e.g. temporal sequence of enzyme development and variation in tissue sensitivity to hormone, Ch. 5). However, light itself does not carry any information specific to particular morphogenic steps. Equally, there is as yet no evidence that specificity in physiological effect derives from any property of the photoreceptor. Therefore, at present, *the specificity of the response is considered to be determined by the specific state of the responding cell.* From this key consideration the concept has been elaborated, particularly by Hans Mohr, that photomorphogenesis comprises two independent stages: pattern specification, in which an endogenous pattern of development is laid down and in which light is not involved; and pattern realization, in which light plays the role of an elective agent, or trigger, which realizes this pre-programmed pattern of development.

[1] See Mohr (1983), Schopfer (1984).

Thus, light initiates many different types of response, not through specification by particular properties either of light or of the photoreceptor, but through different states of competence of the responding cells.

(There is a growing body of evidence from immunochemical studies that there are indeed structurally distinct forms of the phytochrome apoprotein within a plant – see Ch. 4, p. 58. However, the significance of differences in this property of the photoreceptor to differences in response to light is not yet clear.)

The influence of light is usually seen from the very beginning of a plant's development, during seed germination. Further photomorphogenic effects cover an extremely wide range of processes. Basically, a skotomorphogenic mode of development, in which stored energy is utilized almost exclusively for rapid axis elongation, changes to a mode in which light itself is the energy source for growth, differentiation and reproduction: stem extension is slowed, leaf expansion is enhanced, the photosynthetic apparatus is developed, transport pathways (vascular tissues) are elaborated, and timing mechanisms (endogenous rhythms) are initiated.

6.2   GERMINATION[2]

The germination of many types of seed is influenced by light – as can be seen from the sudden appearance of various weed species after soil has been cultivated. There are three kinds of germination response towards light:

(1)   Seeds in which germination is stimulated by light are said to be **positively photoblastic**. As early as 1908, Kinzel noted that out of a total of 974 species, 672 had a light requirement for germination.

(2)   Germination is inhibited by light in **negatively photoblastic** seeds.

(3)   In a relatively few species, germination is unaffected by light. However, in many (seemingly) light neutral seeds, a light response can be induced under certain conditions, such as by ripening or imbibing the seed under unfavourable conditions; 'light neutral' mustard seeds are inhibited by light if germinated under conditions of water stress.

Germination involves growth of the embryonic axis, usually wholly

[2] See Frankland (1981), Frankland & Taylorson (1983), Grime (1981).

by cell enlargement, and is considered to be complete when the radicle emerges throught the seed coat. For an individual seed, it is an all-or-nothing process, but for a seed population the percentage germination reflects a natural variation in response to particular environmental factors. If a seed does not germinate when conditions seem to be favourable, it is said to be **dormant**, and requires some special treatment to initiate germination. If freshly harvested seeds do not germinate, this is termed primary or innate dormancy. When imbibed seeds which are able to germinate are exposed to unfavourable conditions, they often become dormant, a state known as secondary or induced dormancy. The degree of dormancy in a seed population is thus determined not only by genetic factors but also by environmental conditions during both seed production and subsequent imbibition. Light is a major environmental influence at all stages: particular light conditions can induce dormancy in some seeds during ripening or during imbibition; alternatively, light breaks dormancy in many types of seed. Phytochrome is the photoreceptor for all these effects. In general, germination is positively correlated with P_{fr} content; any situation which reduces the level of P_{fr} lowers the percentage germination.

The level of P_{fr} in the mature seed is influenced by the light environment during seed ripening. Daylength is certainly a significant factor – seeds produced under long days are generally more dormant than those ripened under short days – but effects of light quality are more marked and more consistent. Exposure to high levels of far-red radiation during ripening, as in shade or under an incandescent light source, usually imposes dormancy in the form of a light requirement for germination. In this respect, the tissues surrounding the seed also have an effect: if they contain significant amounts of chlorophyll, then the light impinging on the actual seed is relatively enriched in far-red radiation. There is, in fact, a positive correlation between the retention of chlorophyll during ripening and a requirement for light in germination, over a wide range of species (Grime 1981). However, a light requirement for germination can occur even in seeds which ripen while fully exposed to direct sunlight. This is thought to be due to differences in the tolerances of the various phytochrome reactions to the decrease in water content which accompanies seed ripening (e.g. low levels of P_{fr} would result if the process of P_{fr} destruction continued at moisture contents too low to permit photoconversion of P_r to P_{fr}).

Light stimulation of germination occurs in several species after a single inductive exposure to light (e.g. the classical phytochrome experiments with light-requiring varieties of lettuce seed). The sensitivity of a seed to light is directly related to its water content. Generally, 'dry' seeds at a moisture content of around 8% are insensitive to light. However,

R–Fr reversibility of response is usually apparent at a moisture content of around 20% (i.e. at a point well before full imbibition). Within limits, there is reciprocity between fluence rate and exposure time for a standard level of germination. There is a sigmoid relationship between the percentage germination and the log of the fluence (Fig. 6.1). This sigmoid form reflects a 'normal' distribution in degree of light requirement in the population. The fluence required for 50% germination represents the light requirement for the average seed in that population (although this can vary markedly between different seed batches, presumably due to such factors as ripening conditions). If the relationship between fluence and germination response is not symmetrical, this could indicate the presence of more than one mechanism or response to light.

Many positively photoblastic seeds need longer or repeated exposures to light for stimulation of germination. This type of behaviour has occasionally been termed 'photoperiodic', but it is probably not analogous to photoperiodism in, say, flowering. That is, this type of germination response seems simply to reflect a requirement for P_{fr} action over a period of time, rather than interaction of P_{fr} with some endogenous rhythm (Ch. 8). In certain cases, an apparent requirement for long exposures to light may even merely indicate additional effects of phytochrome on increasing proportions of the population.

For *light inhibition* of germination in negatively photoblastic seeds, long exposures to relatively high irradiances are usually necessary, as,

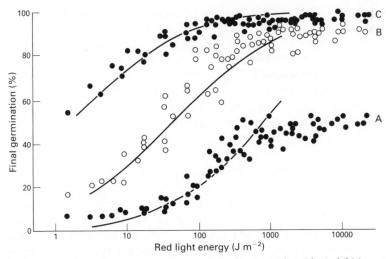

Figure 6.1 Seed germination in relation to light fluence. After 6-hr imbibition at 25°C, batches of *Sinapis arvensis* seed were exposed to red light at fluence rates of 0.3–3.0 W m^{-2}; different seed batches differ in their responsiveness, but all show a sigmoid response to fluence.

for example, in *Avena fatua*, *Nemophila insignis* and *Phacelia tanacetifolia*. The far-red region of the spectrum is generally the most inhibitory, followed by the blue, although in the case of *Phacelia*, red light also inhibits germination. The inhibitory action is thought to operate through the HIR. Therefore, it has been proposed that light has a dual action on germination, with photocontrol operating at two stages in the process:

$$+ve \qquad\qquad\qquad -ve$$

$$\uparrow \qquad\qquad\qquad\quad \uparrow$$

$$P_{fr} \qquad\qquad\qquad HIR$$

Action of P_{fr} in the LER mode promotes germination, while the HIR inhibits germination; differences between positively and negatively photoblastic seeds reflect differences in the importance of these two control points in any particular species.

As with the majority of plant photoresponses, the actual effects of P_{fr} in germination are far from clear. Far-red reversibility of red-light induced germination generally shows a long escape time, suggesting a requirement for P_{fr} over a long time period, rather than activation of a rapid response. Germination involves changes in rates of respiration, in nucleic acid and protein synthesis and in polysome development, but there is no evidence for the direct action of P_{fr} on any of these processes.

Blue light also exerts positive and negative effects on seed germination. The basis for its action is uncertain. The screening effects of seed coverings can confuse the issue of wavelength effectiveness, particularly at shorter wavelengths. A blue-light receptor may be active in some of these responses, but it is generally assumed that blue light affects germination through the action of phytochrome.

The light responses of seeds are markedly affected by other factors, in addition to the ripening environment which has already been mentioned:

(1) Temperature affects the degree of light responsiveness; the differential between light- and dark-germination does not stay constant over a range of temperatures. Furthermore, different temperature pretreatments can alter the subsequent responses to light: chilling often reduces the light requirement of positively photoblastic seeds, while high temperatures increase the requirement for light.

(2) Plant hormones interact in the light responses of seeds; gibberellin can often substitute for an inductive light treatment,

and exogenous cytokinin can sensitize seeds to respond to lower light fluences.

(3) Most generally, the light responses of a seed are completely lost if the covering structures of the seed are removed. This effect of the seed coat could derive from presence of an inhibitor, impairment of O_2 uptake, or mechanical restriction of growth. The natural deterioration of the seed coat under field conditions would thus result in an eventual release from light-dormancy.

The light requirements of a seed population are of ecological significance in several ways. In the first place, they provide a mechanism for the absolute detection of light, which is probably useful in indicating (to the seed) the extent to which it is buried under soil. The spectral environment shows marked qualitative and quantitative changes within a surprisingly shallow depth of soil: in general, at around 1 mm soil depth, the fluence rate is only a few percent of the value at the soil surface and the R : Fr ratio = 0.5 (soil type and water content also influence these values). *Light-requiring* seeds therefore will only germinate when they are on, or very close to, the soil surface, a situation that may have advantages for small seeds with little food reserves. Alternatively, *light-inhibited* seeds will be prevented from germinating near the surface, which may be important for survival during, say, dry conditions. However besides usefulness in absolute light detection, the light requirements also confer on a seed the ability to monitor the spectral quality of an environment. The germination of many *shade-avoiding* species is inhibited by low R : Fr ratios (e.g. under dense canopy), thus ensuring that the actual production of seedlings only occurs under conditions that are suitable for continued growth and development.

6.3 GROWTH OF ETIOLATED SEEDLINGS

6.3.1 *Introduction*

'Growth' is expressed in various ways, which include changes in length, weight, volume and cell number. At the cellular level, it consists of progression through different stages from division through expansion to cell maturation. Cell elongation is the process which accounts for most of the increase in size of a seedling.

Light generally exerts a strong regulatory influence on shoot elongation in seedlings. Phytochrome mediates a major part of this

[3] See Gaba & Black (1983), Taiz (1984).

response, operating through both the LER and the HIR, but there is good evidence that blue-light receptors also affect elongation in most species. Besides variation in the effectiveness of specific wavelengths, there is also great variation in the actual growth responses of different tissues: light can bring about the totally opposite effects of growth stimulation and growth inhibition. Such disparate behaviour is accounted for in an explanation that has come to be known as the Thomson hypothesis: light accelerates all phases of the developmental sequence from cell division through elongation to maturation. This will have the effect of stimulating the growth of younger tissues (through hastening the onset of the elongation phase) but inhibiting the growth of older tissues (through bringing the elongation phase to an end before the full growth potential is expressed). The growth effects obtained by irradiating an etiolated organ at different stages of its development are diagrammed in Figure 6.2. Such a situation has effects not only on the responses of organs of different ages (e.g. red light stimulates the growth of young coleoptiles but inhibits that of older), but also on the responses of different regions within the one organ (e.g. red light stimulates growth in the young apical regions of lettuce hypocotyls but inhibits growth in the older more basal regions).

The idea that the nature of the light–growth response is determined

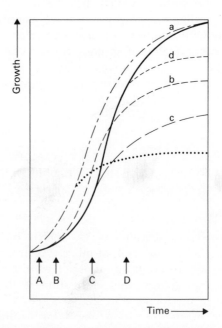

Figure 6.2 Light growth responses and age. The relationship between age at time of exposure (points A–D) and subsequent growth response (lines a–d) of an organ exposed to a single period of irradiation or to continuous irradiation; solid line shows growth in continuous darkness, dotted line in continuous light.

Box 6.1 Growth measurement

In higher plants, features of form, orientation and movement are all based on growth phenomena. The initial stages of many types of investigation involve growth measurement. Of the many measures of size, weight and volume which can be used for this, possibly the most common involve examination of some linear dimension. However, simple, traditional methods based on direct measurement have a number of severe limitations:

(1) they frequently involve manipulation of the plant material, and there are many reports that physical movement itself provokes significant growth responses (an aspect that is also frequently overlooked in other areas of physiological investigation);

(2) such methods quickly become unwieldy with the requirement for more information, which involves more individual measurements in both time and space;

(3) most important, they suffer from a considerable lack of sensitivity, again in both time and space (i.e. they cannot detect fluctuations in growth rate or rapid responses to changes in the light environment).

An indirect means of measurement which has been used to great effect in the investigation of light–growth reactions, involves a *linear displacement transducer* (Cosgrove 1981, Morgan *et al.* 1981). There are various forms of this device, but in general a plant part is attached to one end of a pivoted lever or some form of position transducer (Fig. 6.3). As the plant grows, movement of the lever or generation of a voltage in the position transducer provides a continuous record of growth. The great advantage of this method is that changes in growth rate are recorded as they actually happen. The method has been used particularly to study changes in growth in simulated sunfleck conditions, and to investigate the rapid growth responses to blue light.

However, not only do growth rates fluctuate temporally, growth patterns change spatially. A limitation of linear transduction methods is that they only record total growth, not the region of growth nor change in the distribution of growth. (A linear transducer could miss the significant growth stimulation and inhibition which can be simultaneously induced by light in different regions of an organ.)

Techniques of *time-lapse photography* provide information in both temporal and spatial dimensions, with a considerable degree of resolution. Plant material carrying recognizable marks is filmed and, subsequently, the displacement of the marks due to growth can be measured frame by frame. (Lanolin 'hairs' or appropriate beads can be used to mark the plant; see below, Fig. 6.6; at bead spacings of 1 mm, growth increments of 50 μm can be readily detected.) A recent innovation in this area involves the use of video equipment to record growth (Gordon *et al.* 1984). This avoids the delay and expense of film processing, and also (a point of particular relevance to the study of light responses), allows growth to be recorded under conditions of irradiation ($\lambda > 800$ nm) to which the plant is insensitive.

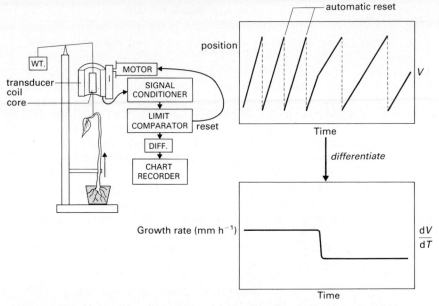

Figure 6.3 One type of continuous growth-measuring apparatus. Growth of the seedling raises the attached transducer core, generating a voltage proportional to position; this voltage is electronically differentiated to give a direct record of growth rate; a reset circuit maintains the transducer core within its linear range.

by the stage of development of an organ is a key concept for the interpretation of many of the effects of light on seedling growth. Such a pattern of response could be considered a further manifestation of the way in which *phytochrome elicits response from a preprogrammed pattern of development*. It is particularly characteristic of responses to brief, inductive exposures to red light, but it is not yet clear whether the growth responses specifically induced by blue light also depend upon the physiological state of the tissue. (Growth responses to blue light, and in the HIR, seem to be more generally inhibitory.)

Actual methods of measuring growth in seedlings are discussed in Box 6.1.

6.3.2 Monocotyledons[4]

In monocotyledons, the organs whose light–growth responses have received most attention are the coleoptile and the mesocotyl of the grass seedling (Fig. 6.4). The coleoptile acts as a protective sheath around the apical meristem and leaf primordia of the shoot, during upward growth through the soil. Increase in size of the coleoptile is

[4] See Gaba & Black (1983), Mandoli & Briggs (1981), Schaer *et al.* (1983).

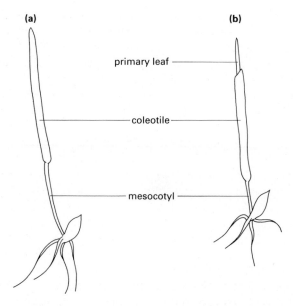

(a)   (b)

primary leaf

coleotile

mesocotyl

Figure 6.4 Differences in morphology in 5-day old oat seedlings: (*a*) grown in continuous darkness; (*b*) exposed to white light at four days.

due mainly to cell extension, and in darkness, the coleoptile grows much more than the enclosed shoot axis (the deeper a grass seed is buried, the longer the coleoptile will be). When the seedling is exposed to light, the relative rates of coleoptile and shoot growth change (to the extent that, under daylight conditions, the true primary leaves of the shoot eventually emerge through the coleoptile tip).

Details of the coleoptile's light responses are a function not only of the species, but also of the age (size) of the organ and the quality and duration of irradiation. Limited irradiation treatments ($< 10^5$ J m^{-2} red, blue or white light) generally tend to stimulate the growth of young (short) coleoptiles, and to inhibit the growth of older (long) coleoptiles. (Within a single coleoptile, this is the type of light treatment which can simultaneously stimulate growth in apical regions and inhibit growth in more basal regions, so that total growth is unchanged.) Thus, the responses of the coleoptile to limited light exposures are a function of the physiological state of the tissue, and are compatible with the Thomson hypothesis. Exposure to long-term irradiation, particularly with blue or white light, simply results in a general inhibition of coleoptile growth.

The mesocotyl is the transition zone between the base of the coleoptile and the attachment of the plumular tissue to the scutellum, and is often equated with a first internode. Growth in this region comprises both cell division and enlargement, and can be very extensive in dark-grown seedlings (Fig. 6.4). However, growth of this organ is extremely sensitive to inhibition by light, with both cell division and cell elongation being affected; cell division seems to be the more sensitive process. Light inhibition of growth in the mesocotyl is one of the most sensitive photomorphogenic responses: the mesocotyls of dark-grown seedlings show growth inhibition by light levels equivalent to three minutes of moonlight. Such routinely grown 'dark' seedlings have usually been exposed to dim green safelights, at least during watering and manipulation. In seedlings grown in complete darkness, so-called 'analar darkness', light effects on mesocotyl growth have been recorded (Mandoli & Briggs 1981) at fluences four orders of magnitude lower still (very-low-fluence-response). (Over 20 years ago, after much painstaking investigation of mesocotyl growth in the absolute darkness of London cellars, the plant physiologist C. L. Mer pointed out that there was no such thing as a 'safelight' for plant growth.)

Thus the coleoptile and mesocotyl show opposite growth responses to limited exposures at low levels of red light (Fig. 6.5). The process of cell elongation itself also seems to be oppositely affected in each organ (stimulated in the coleoptile, but inhibited in the mesocotyl). These responses perhaps suggest that the Thomson hypothesis is too simplistic, in its original form, to account for every situation (Schaer et al. 1983).

The influence of light on specific aspects of cell extension is considered in Box 6.2.

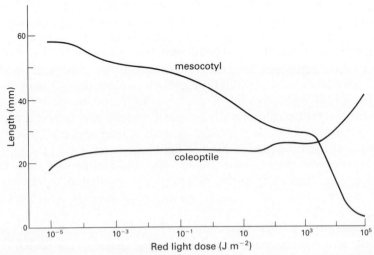

Figure 6.5 Relationships between growth and red light fluence in the coleoptile and mesocotyl of oat.

Box 6.2 Cell extension

The rate of cell extension ultimately derives from the balance between the yielding properties and extensibility of the cell wall and the ability of the cell to take up water from its surroundings (Cosgrove 1983, Gaba & Black 1983, Taiz 1984). The relationship is formally represented in the equation:

$$\text{growth rate} = \frac{L_p.W_{ex}(\Psi_{ext} - \Psi_{int}) - Y}{L_p + W_{ex}}$$

where L_p = hydraulic conductivity, W_{ex} = wall extensibility, Y = wall yield stress, Ψ_{ext} = external water potential, and Ψ_{int} = internal water potential. Change in any one of these factors could theoretically result in a change in growth rate (e.g. increased growth could arise from increases in L_p, W_{ex} or Ψ_{ex}, and decreases in Y or Ψ_{int}.

A number of observations indicate that light-induced changes in growth rate are correlated with changes in wall extensibility: growth inhibition in etiolated pea stems in red light is associated with a decrease in wall plasticity; conversely, stimulation of apical growth in coleoptiles by red light is accompanied by an increase in wall plasticity. Inhibition of growth in blue light also involves a decrease in wall loosening. Work with auxin has provided considerable evidence that changes in wall extensibility, and hence growth, result from changes in the rate of proton extrusion from the cell membrane. Again, stimulatory and inhibitory effects of red light on growth have been correlated with, respectively, increased and decreased wall acidification.

Of course, in the absence of details concerning the primary action of any photomorphogenic receptor, consideration of the actual mechanisms by which light affects cell growth must remain fairly speculative. However, it is intriguing to think that the rapid effects of phytochrome (and of blue-light receptors) on certain membrane processes may in some way be connected with the changes in wall properties which accompany changes in growth rates.

6.3.3 Dicotyledons[5]

Growth of the plumule occupies a central role in the events leading to seedling establishment, particularly with small-seeded species where limited food reserves mean that survival is dependent upon getting the potentially photosynthetic tissues into light as quickly as possible.

The plumular hook of the dicotyledonous seedling ensures that the delicate meristematic primordia of the apex are protected as they are pulled, rather than pushed, upwards through the soil (Fig. 6.6). (Virtually all dicot organs involved in upward growth through soil, do

[5] See Gaba & Black (1983), MacDonald *et al.* (1983).

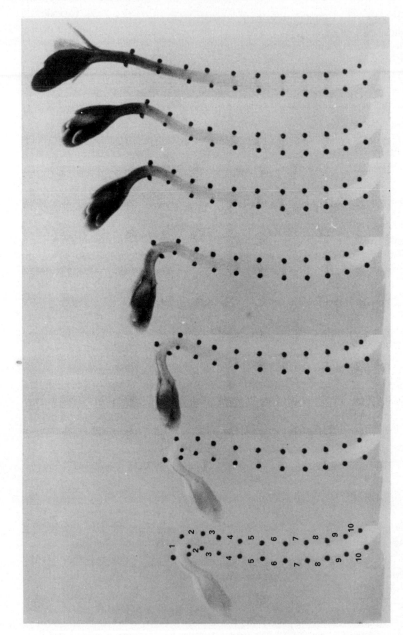

Figure 6.6 Time-lapse sequence of growth at 3-hourly intervals in a cress seedling, showing the hypocotyl hook configuration in darkness (extreme left) and hook unfolding during subsequent growth in 18 hrs white light. (The marker beads allow growth in different parts of the organ to be measured.)

so with the apical region in the form of an arch or hook.) The hook is not preformed within the seed, but arises during early stages of seedling growth in response to environmental signals: hook formation in lettuce is hastened by brief exposure to light, whereas in many other seedlings, gravity plays a major role in hook formation and mainten-ance, through positive gravitropism in the apical regions (the remainder of the organ is negatively gravitropic.) In some types of seedling, such as *Lepidium sativum* (cress), growth within the actual hook region is minimal, the major upward thrust of the seedling being sustained by cell extension in the shank (Fig. 6.6, zones 3–4). In other species, such as *Phaseolus vulgaris* (French or kidney bean), growth does occur within the hook, particularly on the outer (convex) side of the apical half of the hook and on the inner (concave) side of the basal half – that is, the morphological form of the hook is maintained, though its component cells continually change; the hook literally 'flows' upwards.

When the hook reaches the soil surface, changes in these patterns of growth bring about hook opening or unfolding. Exposure to light results in marked growth stimulation on the inner concave side of the hook (Fig. 6.6, zones 1–2), while growth of zones on the outer convex side is unaffected or even inhibited. Light perception for these responses occurs within the hook region itself. In most species, hook opening is induced by the LER (phytochrome), although the process generally occurs much more rapidly in continuous white light (suggesting that the HIR or a blue-light receptor is also involved); lettuce is unusual in that hook formation is induced by the LER and opening only by the HIR. The different growth responses initiated by light during hook opening (i.e. stimulation and inhibition) serve as another example in which the effects of light are related to the physiological status of the responding cells. Indeed, here too the responses can be interpreted in terms of the Thomson hypothesis: the immature cells on the inner side of the hook are accelerated into their expansion phase, while the previously growing cells of the outer side are accelerated into their maturation phase.

During the growth of a young shoot, regions showing the greatest rates of elongation are generally confined to the apical half of the plumule. However, the actual growth patterns differ in etiolated and green seedlings of the same species. For example, in etiolated cress seedlings, the growth rate is high throughout the whole apical half of the hypocotyl; in de-etiolated seedlings, growth is more restricted to the apical quarter of the organ. That is, the slower total growth of green seedlings (0.5 mm h^{-1} in comparison to 1.0 mm h^{-1} in etiolated seedlings) derives not so much from the slower rate of a process, but from the smaller region of the organ that is actually growing. In

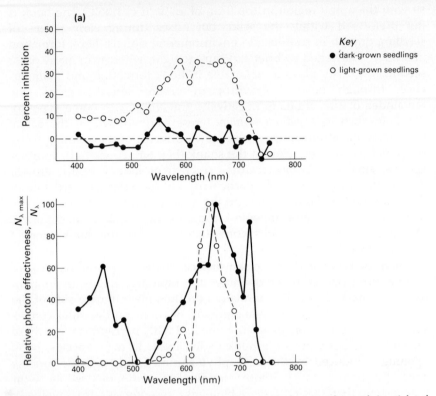

Figure 6.7 Effects of different types of light treatment on hypocotyl growth in etiolated and green mustard seedlings at 54–78 hours after sowing: (*a*) a 30-minute 'end-of-day' irradiation treatment at different wavelengths, followed by 23.5 hours of darkness; (*b*) 24 hours continuous irradiation at different wavelengths.

continuous white light, growth is even more confined to discrete apical zones. These patterns of growth again seem compatible with the Thomson hypothesis.

Photocontrol of seedling growth in most species is mediated through both phytochrome and blue-light receptors. In etiolated material, action of the LER on growth is, in general, only weakly apparent, and continuous or repeated exposures to light are necessary in order to achieve significant growth inhibition (Fig. 6.7). The standard regions of far-red, red and blue are generally the most effective (although in lettuce and cucumber seedlings, sensitivity to far-red irradiation can be lost with age). De-etiolated seedlings are much more sensitive to LER-exposures, but also have lost sensitivity to HIR-Fr. These changes after de-etiolation are discussed in the next section.

Inhibition of growth by HIR-B light is strongly apparent in both etiolated and green seedlings of most species. Much of the evidence for the existence of a non-phytochrome receptor for blue light in higher plants is derived from studies of hypocotyl growth. This evidence (Mohr 1984, Koorneef *et al*. 1980, Schäfer & Haupt 1983, Thomas 1981) is described in Ch. 4; it includes differences between red and blue light in regard to:

(1) relative effectiveness, at similar values of φ (phi, symbol for phytochrome photoequilibrium);
(2) timing of responses to light treatments;
(3) regions of light perception;
(4) interaction of light treatments with hormones;
(5) responses of growth mutants to light treatments.

The effects of blue light on growth are imposed very rapidly, usually within less than five minutes from the onset of irradiation; in etiolated mustard seedlings, exposure to just 11 W m^{-2} brings about a decrease in growth rate within 40–60 s (Fig. 4.11). (Obviously, such responses can only be measured by methods which provide a continuous record of growth.) Usually, there is also rapid recovery of growth, within about 20 minutes from the end of the exposure to blue light. (Specificity of blue light action is not completely distinguishable by speed of response, however, since green shoots which have had a light pretreatment can also show phytochrome-mediated changes of growth rate within minutes – see next section.) The effects of blue light are

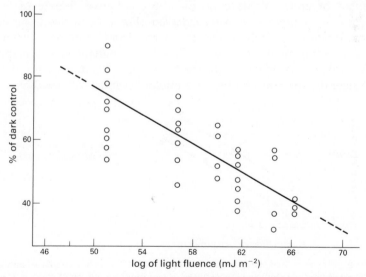

Figure 6.8 Growth response of cucumber hypocotyls as a function of blue light fluence. Blue light at fluence rates of 0.4–10 W m^{-2} was given for periods of 5 min. except at the lowest dose where 30 sec. exposures were used.

highly dependent on fluence rate, and there is a linear relationship between response and the log of the light dose (Fig. 6.8). (Such a log–linear relationship is characteristic of many photobiological responses, and' is sometimes referred to as the *Weber–Fechner law*; its basis is uncertain, but it is thought to derive from a 'logarithmic amplification' of the light signal rather than from any feature of the photoreceptor itself – see Schäfer & Fukshansky 1984.) Finally, rapid blue-light induced changes in growth rate occur not only as a result of transfer from darkness to light, but also in response to sudden increases in fluence rate (Cosgrove 1981, Gaba & Black 1983).

6.4 GROWTH OF GREEN TISSUES

6.4.1 Introduction[6]

The light growth responses of de-etiolated plants are quite different from those of etiolated seedlings, in both quantitative and qualitative terms. Quantitatively, the growth of green tissues is much more sensitive to light. This is seen in two respects: first, growth in green seedlings is more strongly affected by relatively short 'end-of-day' irradiation treatments (Fig. 6.7); and secondly, under continuous (red) light, growth in green seedlings is more sensitive to lower fluence rates (Fig. 6.9). However, the most marked change in the light growth responses of green plants concerns the qualitative change of *loss of sensitivity to continuous far-red irradiation* (Fig. 6.7). Indeed, far from being inhibited by high levels of far-red irradiation, the extension growth of some plants seems actually to be 'enhanced' – a response that has major consequences for the behaviour of certain types of plant under conditions of shade. (De-etiolated mustard seedlings are

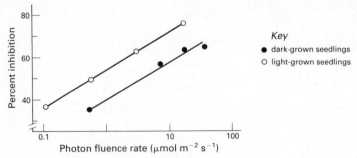

Figure 6.9 Effects of increasing fluence rate on hypocotyl growth in etiolated and green mustard seedlings; seedlings were exposed to continuous red irradiation (λ 656 nm) during 54–78 hr after sowing.

[6] See Jabben & Holmes (1983).

118

relatively unusual in also showing a loss of sensitivity to the blue region.)

The characteristics of phytochrome also differ in de-etiolated tissues (this has been uncovered mainly through investigation of Norflurazon-treated, chlorophyll-free seedlings). Phytochrome is present in de-etiolated seedlings at levels approximately 50 times lower than those in etiolated seedlings. However, it seems to be much more stable in de-etiolated tissues: there is no dark reversion, and physiological dark destruction of P_{fr} occurs much more slowly. It is not clear whether the lower levels of phytochrome in de-etiolated tissues result from a new equilibrium between processes of synthesis and destruction being set up, or from the appearance of a different (stable) pool of phytochrome – perhaps even the product of different phytochrome genes. The extent to which such differences in the characteristics of phytochrome account for the differences in light-responsiveness between etiolated and green tissues is also not entirely clear. It has been suggested that the loss of HIR–Fr activity in green tissues may be due to the loss of competition between P_{fr}-producing light reactions and the P_{fr}-removing dark reactions.

These general differences in growth responses and phytochrome behaviour between etiolated and green tissues are summarized in Table 6.1.

6.4.2 Light growth responses[7]

Various different types of light treatment have been used in experimentation with de-etiolated plants. *End-of-day* treatments consist

Table 6.1 Differences between etiolated and green seedlings in light–growth responses and phytochrome characteristics.

Seedling type	Responsiveness to irradiation		Phytochrome characteristics
	End-of-day treatments	Continous irradiation	
etiolated	poor responses	Fr, R, B effective; high fluence required; slow (20-min) response to R	'high' concentrations $(2 \times 10^{-7}$ mol dm$^{-3})$ labile (extensive dark destruction of P_{fr})
de-etiolated	sensitive responses	Fr ineffective; sensitive to low fluence; fast (5-min) response to R	'low' concentrations $(2 \times 10^{-9}$ mol dm$^{-3})$ stable (slower dark destruction of P_{fr})

[7] See Jabben & Holmes (1983), Morgan *et al.* (1981), Smith & Morgan (1983).

of exposing the plant material to particular light conditions at the end of a standard day or photoperiod. The stimulation of stem extension by far-red irradiation at the end of the main photoperiod provided one of the first demonstrations (in 1957) that phytochrome is involved in the photocontrol of growth in green plants; the effect is negated by an immediate treatment with red light. (This serves to emphasize that, overall, P_{fr} *tends to inhibit stem growth* – a statement that should be kept clearly in mind for subsequent elaboration.) Although end-of-day changes in the R : Fr ratio do occur in nature, their limited extent and duration in relation to other aspects of the spectral environment probably means that they are not a major factor in the natural photocontrol of growth.

Other types of treatment that involve only limited light exposures also influence the growth of green tissues. Changes in growth rate within a minute of exposure to blue light, of course, characteristically indicate the involvement of a blue-light receptor. But in green seedlings, treatments with red or far-red light also induce fairly rapid growth responses: exposure of *Chenopodium album* (fat hen) to far-red light results in an increased rate of stem extension within seven minutes; this response too can be negated by subsequent red light treatment. If it is only the leaves that are exposed to far-red irradiation, growth responses still occur in the stem, albeit several hours later. This suggests that phytochrome induces some kind of transmissible effect (in contrast to the more localized action of a blue-light receptor). These rapid responses to limited light treatments probably are of significance to certain situations in the natural environment, for example, in relation to behaviour in sunflecks under vegetation canopies.

Growth responses to extended light exposures are also determined by the relative level of P_{fr}. Again in *Chenopodium*, the log of stem extension rate shows an inverse linear relationship to the level of the phytochrome photoequilibrium (Fig. 6.10a). Growth rate is therefore influenced by the *light quality*, as manifest in the R : Fr ratio (Fig. 6.10b):

$$\text{low R : Fr} \rightarrow \text{low } \phi \equiv \text{low } P_{fr} \rightarrow \text{high growth rate}$$

(Note that in etiolated seedlings, long exposures to high levels of far-red irradiation would inhibit growth through the HIR.) In treatments with continuous light, growth is also influenced by *light quantity*, in the sense that lower fluence rates allow greater stem or hypocotyl extension (Fig. 6.9).

It is possibly these types of extended light treatments that are of most relevance to an understanding of the degree to which light influences plant extension growth in the natural environment. For

(a)

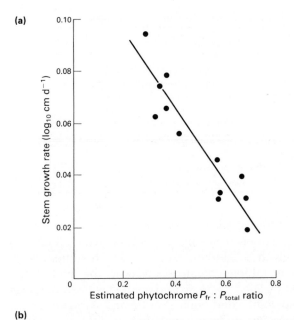

(b)

Figure 6.10 Effects of light quality on stem extension in *Chenopodium album*. (*a*) Rate of stem growth at different phytochrome photoequilibrium values. (*b*) Seedlings grown for 14 days in a range of simulated shadelight treatments at different R : Fr (ζ) values; $\zeta = 2.28$ is the control treatment of white fluorescent light; $\zeta = 0.18$ simulates very deep shade.

Table 6.2 Plant responses to shade conditions.

Plant type	Responses
shade-avoider	enhanced axis elongation
	increased internode extension
	increased petiole extension
	strong apical dominance
	limited branching
	limited leaf expansion
shade-tolerator	leaf development
	slow growth
	low respiration

example, when plants of *Chenopodium album* were grown in simulated shadelight conditions of enriched far-red irradiation (which increases stem extension), and given a 30-minute end-of-day treatment with fluorescent light (which inhibits stem extension), the end-of-day treatment only inhibited by 20% the increased stem extension that was attributable to the shadelight conditions (Smith and Morgan 1983). That is, at least in *Chenopodium* under simulated shadelight, it seems to be the whole daytime irradiation period that exerts the major influence on stem extension, rather than the end-of-day conditions.

6.4.3 *Growth in the natural light environment*[8]

The largest changes in the light environments of terrestrial habitats are found under vegetation canopies, where there are large decreases both in the R : Fr ratios and in the fluence rates. Different types of plants have evolved different ways of responding to this kind of situation (Table 6.2). *Shade-avoiders* show enhanced axis development in response to shade conditions – increased internode and petiole extension, strong apical dominance (little branching), and limited leaf development. This type of reaction to shade is seen in competitive species found normally in open environments, such as *Chenopodium* (fat hen), *Senecio* (ragwort) and *Chamaenerion* (fireweed). On the other hand, *shade tolerators* tend to show greater leaf development. Generally, these species also tend to be slow growing and to have low respiration rates, low photosynthetic light-compensation points and low rates of leaf turn-over (i.e. they show modes of development which conserve resources). Significantly, shade-tolerators like *Mercurialis perennis* (dog's mercury) show little

[8] See Grime (1981), Holmes (1981, 1983), Morgan (1981), H. Smith (1982, 1983a), Smith & Morgan (1983).

response to the removal of shade. It is particularly the shade-avoiding species which show large growth reactions to changes in the light environment.

There is considerable evidence, mainly from the work of Harry Smith and his colleagues, that phytochrome is the detector for the changes in light quality in shade conditions:

(1) The R : Fr ratio ζ determines the phytochrome photoequilibrium ϕ, with the greatest responsiveness of ϕ over precisely the range of ζ found in shade conditions (Fig. 4.8).

(2) The extent of plant response shows a specific relationship to the level of phytochrome photoequilibrium (Fig. 6.10a);

(3) Changes in the R : Fr ratio result in changes in growth rate, particularly in shade-avoiding species (Fig. 6.10b).

However, changes in light quantity also produce changes in growth rate, again particularly in shade-avoiding species. (This cannot be due simply to photosynthetic effects, since under conditions of low fluence rate, where photosynthate must be limiting, stem extension is greater.) Both phytochrome and blue-light receptors are involved in the detection of light quantity, although their relative importance in this respect seems to differ between species. For example, green seedlings of *Sinapis alba* are virtually insensitive to blue irradiation (phytochrome must therefore be responsible for responses to light quantity); *Chenopodium rubrum*, however, is much more sensitive to blue than to red irradiation.

It is not clear whether plants utilize light quality or light quantity as the more generally significant indicator of shade conditions (Holmes 1983). It seems to depend upon the particular situation of both tissue state and actual spectral environment. For example, in etiolated mustard seedlings, growth shows a greater response to change in light quantity than to change in the R : Fr ratio, while in green seedlings growth is highly responsive to change in both light quantity and quality (i.e. the green seedling could potentially distinguish between shading by inanimate objects and by particular types of vegetation canopies). However, even in the growth of green seedlings, the relative effects of light quality and light quantity can differ in different situations. For example, under light of very low R : Fr ratio (i.e. conditions analogous to deep vegetation shade), hypocotyl extension in cucumber seedlings is more responsive to change in fluence rate than to change in R : Fr ratio; but at higher R : Fr ratios, growth is responsive to change in both fluence rate and R : Fr ratio.

6.5 DE-ETIOLATION AND LEAF EXPANSION

When a higher plant first comes into contact with light, an obvious development is that parts of it become green. However, despite the fact that greening is such a conspicuous process, and one which is crucial to the continued survival of the plant, there is still a considerable lack of information concerning many of its aspects. The reason for this perhaps surprising state of affairs is that transformation into a photoautotrophic organism encompasses such a multiplicity of processes which occur at different levels of cellular organisation and in different cellular compartments. (Also, details of timing and regulation of the greening process vary between situations, e.g. between etiolated seedlings, mature aerial tissues and root tissues.)

In this section, a general outline is given of the types of process which occur during greening and the points at which light exerts regulatory effects. Events are treated at increasing levels of cellular organisation, namely:

  (1) at the molecular level (e.g. chlorophyll synthesis);
  (2) at the ultrastructural level (e.g. chloroplast development);
  (3) at the anatomical level (e.g. aspects of leaf development).

6.5.2 Chlorophyll synthesis[9]

The general sequence of events in chlorophyll synthesis is shown in Figure 6.11 as five stages, all of which occur in the developing chloroplast.

  (1) *ALA synthesis.* The first so-called committed metabolite in the pathway is δ-aminolevulinic acid (ALA) – the earliest metabolite unequivocally identified as a precursor of chlorophyll. However, synthesis of ALA is still one of the least understood and most controversial stages in the pathway. In haem synthesis in animals, ALA is produced by the condensation of succinyl-CoA with glycine, through the action of the enzyme ALA-synthetase. This enzyme has not been generally detected in plant tissues. In higher plants, ALA is thought to originate from a 5-carbon precursor, glutamic acid or α-ketoglutaric acid, through reactions termed, respectively, the 'GSA-path' and the 'DOVA-path'. The production of ALA is a light-regulated stage.
  (2) *PBG formation.* Two ALA molecules condense to form the

[9] See Kasemir (1983), Castelfranco & Beale (1983).

porphobilinogen ring structure (PBG). PGB is the first pyrrole intermediate in the pathway, and its formation marks the start of the assembly of the tetrapyrrole skeleton.

(3) *Protoporphyrin formation*. Four PBG units initially form an unstable linear tetrapyrrole which is then enzymatically closed to give a cyclic porphyrinogen molecule. A series of reactions then produces the cyclic tetrapyrrole, protoporphyrin, which is the last precursor common to the synthesis of haems, bilins and chlorophylls.

(4) *Protochlorophyllide formation*. The chlorophyll branch of the pathway begins with the insertion of magnesium into the protoporphyrin nucleus. A further series of reactions, the details of which are not entirely clear, produces protochlorophyllide, which now contains the fifth, isocyclic ring of the chlorophylls. By this stage also, the molecule is conjugated to a protein. This chromophore-protein complex is called protochlorophyllide-holochrome, and this is the form which is present, to a limited extent, in etiolated tissues.

(5) *Chlorophyll a production*. Protochlorophyllide-holochrome is reduced to chlorophyllide-protein (saturation of the double bond between C7 and C8 of ring D). In most angiosperms, this reduction can only be carried out when the chromophore is in a photoexcited state – a second light-controlled step in chlorophyll synthesis, and protochlorophyllide itself is the photoreceptor (λ_{max} = 650 nm). This photoreductive step is one reason why the greening process in angiosperms requires *continuous exposure* to substantial amounts of light. The reduction also requires that protochlorophyllide be complexed to its apoprotein; many workers ascribe an enzymatic role to the protein itself and replace the term 'holochrome' with the more descriptive name, NADPH-protochlorophyllide-oxidoreductase. (In gymnosperms, this reductive step can be carried out wholly enzymatically, such that conifers, say, can form a limited amount of chlorophyll in complete darkness.) The final stages of chlorophyll synthesis are concerned with the addition of the long-chain phytyl tail to chlorophyllide. Other chlorophylls are derived from chlorophyll *a*.

The overall regulation of this massive biosynthetic sequence is complex, even when consideration is restricted to aspects of photocontrol. Exposure of etiolated tissues to continuous light results in three consecutive phases of chlorophyll build-up: an initial rapid phase of pigment appearance, a lag phase and a final steady phase of pigment accumulation.

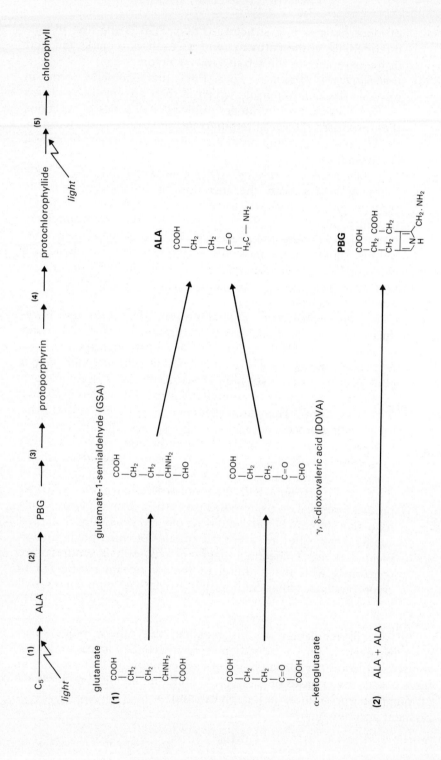

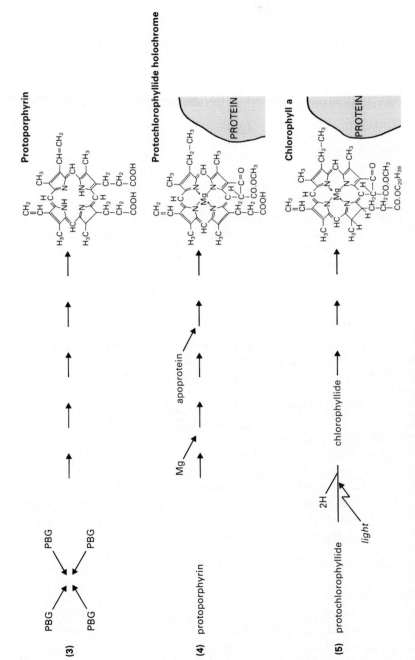

Figure 6.11 General sequence of events in chlorophyll synthesis.

The initial rapid phase consists of the immediate photoconversion of the small amounts of protochlorophyllide which are minimally present in etiolated angiosperm tissues. During this process a series of absorption changes occur in the tissue. These probably derive not only from chemical changes in the chromophore, but also from organizational adjustments of the pigment-protein complexes on the chloroplast membranes:

$$\text{Protochllde}_{650} \xrightarrow{\lambda\,650\text{ nm}} \text{Chllde}_{678} \xrightarrow{} C_{684} \xrightarrow{} C_{672} \xrightarrow{} \text{etc.}$$

In the subsequent lag phase, regulatory effects of light on other stages of chlorophyll synthesis can be demonstrated. Two processes in particular show distinct R–Fr reversibility of response. The rate of ALA synthesis is markedly enhanced by red light, probably by an effect of P_{fr} on production of the enzyme responsible for ALA synthesis; and the synthesis of chlorophyll-apoprotein is induced by red light. Note that these two processes, ALA synthesis and chlorophyll-apoprotein synthesis, occur in the chloroplast and in the cytoplasm, respectively. That is, phytochrome is involved in the co-ordination of molecular events that take place in separate cellular compartments.

6.5.3   Chloroplast development[10]

The generic term **plastid** describes a class of organelles unique to plants. They are distinguished by a double bounding membrane and an internal fretwork of membranes and plastoglobuli (lipid droplets). Plastids are characterized on the basis of their pigment content and internal structure. Ultrastructural investigations suggest a degree of ontogeny between different types of plastid, a generalized scheme of which is shown in Figure 6.12.

Proplastids, with few internal membranes or vesicles, are the simplest form of plastid, and are found in meristematic regions in very young or dividing cells. (The term probably covers a heterogeneous collection of organelles, only some of which have the capacity to develop into chloroplasts). *Etioplasts* are larger plastids found in dark-grown tissues. They characteristically contain one or more prolamellar bodies (Fig. 6.13), paracrystalline structures that consist mainly of the lipid components of membranes. Limited amounts of protochlorophyllide are present, as are small amounts of the photosynthetic machinery, e.g. the redox coenzymes and the carboxylation enzyme RUBISCO (ribulose-bisphosphate-carboxylase-oxygenase); exposure to light initiates the protochlorophyllide photoconversion and also results in

[10] See Bjorn (1980), Jenkins *et al.* (1983), Lichtenthaler *et al.* (1980), Possingham (1980), Schopfer & Apel (1983), Virgin & Egneus (1983).

massive further synthesis of these components. *Chloroplasts* (Fig. 6.13) are plastids whose internal membranes carry chlorophyll. *Chromoplasts* are usually much larger, with a very high content of membranes and carotenoids; they are considered a degenerative form of plastid, and are characteristic of flower and fruit tissues. Other types of plastid include *amyloplasts* (high starch content, found in storage tissues) and *leucoplasts* (colourless plastids of epidermal and storage tissue).

When dark-grown aerial organs are exposed to light, the etioplasts show almost immediate ultrastructural change: within seconds, the prolamellar body breaks up to form vesicles and tubules. This initial event may be associated with the photoconversion of existing protochlorophyllide. If irradiation is continued over the next few hours, more membranes appear, and eventually become organized into the familiar grana-stacks of the chloroplast. These events occur during the steady, third phase of chlorophyll accumulation under high irradiances. It is during this stage that exposure to high-irradiance blue light results in the development of sun-type chloroplasts (Fig. 6.14): smaller chloroplasts with few grana but a very high content of P700, redox coenzymes and RUBISCO. These effects of blue light are only weakly apparent in the phototransformation of etiolated seedlings, but are strongly present in mature tissues. Again, in lower organisms, and in roots, exposure to blue light is absolutely necessary for initial transformation of the prolamellar body (Bjorn 1980, Lichtenthaler *et al.* 1980).

Thus, during greening, many events occur at various levels of organization (Fig. 6.15). Three photoreceptors are involved in the light regulation of the processes: protochlorophyllide, phytochrome and a blue-light receptor. Protochlorophyllide seems to play a key regulatory

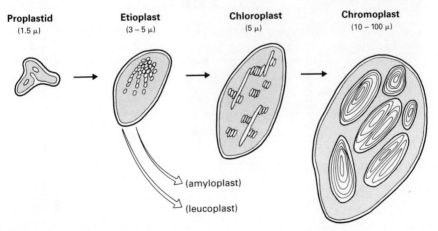

Proplastid	Etioplast	Chloroplast	Chromoplast
(1.5 μ)	(3 – 5 μ)	(5 μ)	(10 – 100 μ)

(amyloplast)

(leucoplast)

Figure 6.12 Developmental relationships between various types of plastid.

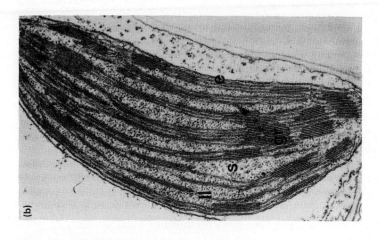

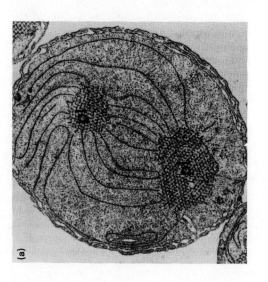

Figure 6.13 Electron micrographs of two types of plastid: (a) etioplast containing two paracrystalline prolamellar bodies (pl) and a rudimentary internal membrane system (× 15 000, micrograph by R. M. Leech; (b) chloroplast showing bounding envelope (e), stroma (s), granum (gr) and inter-granal membranes (ll) (× 10 000, micrograph by W. W. THomson).

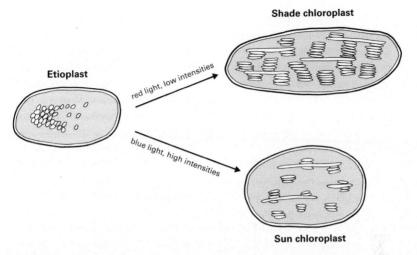

Figure 6.14 Effects of light on the formation of two types of chloroplast differing in photosynthetic activity and ultrastructure; sun chloroplasts are formed at high light intensities and in blue light.

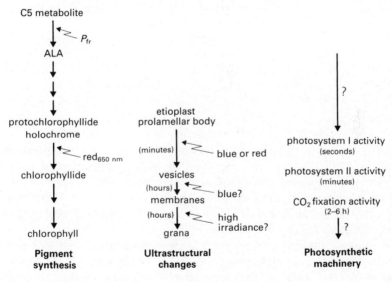

Figure 6.15 Summary of the general events during (photo-)transformation of an etioplast to a chloroplast.

rôle, but *not* by acting as a photomorphogenic 'sensor' pigment. The presence of protochlorophyllide seems to signal a block to pigment synthesis, to apoprotein synthesis and to ultrastructural development. The removal of protochlorophyllide is actually powered by light, and must occur continuously throughout the complete greening process.

131

6.5.4 Leaf expansion[11]

The details of leaf development vary greatly between species. As a crude approximation, cell division determines leaf shape, while cell expansion determines leaf size (as much as 80% of final leaf size can be attributed to cell expansion). Cell division does not stop uniformly throughout the leaf: it ceases first in the distal regions and later in the proximal regions. Within this general pattern, there are particular distributions of local meristems whose continued differential activities account for particular characteristics of leaf shape and anatomy. Light affects both cell division and elongation, and thus affects both leaf size and shape.

At first glance, light may seem to exert opposite effects on stems and leaves, in being generally stimulatory towards leaf growth. Again, however, these different modes of behaviour can be accommodated in terms of the light-realization of developmental pattern – P_{fr} evokes responses specific to the particular states of competence of the responding cells. (In terms of the Thomson hypothesis of light acceleration of the sequence of cell development, it may be that, in darkness, leaf cells never develop beyond a 'juvenile phase'; therefore, light inhibition of leaf growth is never manifest.)

Most aspects of leaf development are markedly influenced by light. Total leaf size is generally 20-fold greater in light than in darkness. This effect is in large measure a response to light quantity, which can be expressed in two ways: total fluence, or fluence rate. Leaf *area* shows a good correlation with *total* fluence received, between the values of 0.4–4.0 MJ m^{-2} (i.e. outside this range, expansion is minimal at very low or very high fluence). However, leaf *thickness* shows a better correlation with fluence *rate*; greater thickness derives both from increased cell size and cell numbers, particularly number of palisade layers. Light quality can also exert a morphogenic influence, as in *Taraxacum officinale* (dandelion), where short end-of-day treatments with far-red radiation result in the production of smooth-edged, rather than serrated, leaves (Fig. 1.4).

A major influence of light on leaf development is seen in **sun** and **shade leaves**. The expression of these syndromes is regulated at various levels. That is, they can be characteristic of distinct species which occupy open or shaded habitats; within this level, there are obligate sun or shade species and also phenotypically adaptable species. Alternatively, at the level of the individual organism, leaves of the same plant can show sun or shade characteristics at different positions in a canopy. However, even in the case of an adaptable individual or species, most characteristics of particular leaves are fixed

[11] See Gaba & Black (1983).

during leaf development; there can be little change in leaf characteristics even if the light conditions change markedly. (The sudden change in light conditions between a well lit greenhouse and a domestic environment can be a cause of leaf-drop, followed, sometimes, by the development of new leaves with characteristics more appropriate to the new light environment.)

Anatomical differences between sun and shade leaves are illustrated in Figure 6.16. Generally, shade leaves are broader and thinner, with fewer layers of palisade mesophyll, shorter palisade cells and a more dispersed spongy mesophyll. Differences between sun and shade leaves, in this aspect of size, are often quantitatively expressed as the ratio of mesophyll area : leaf area (A_{mes} : A). This ratio of 'internal area : external area' would thus be low (13) for shade leaves and high (40) for sun leaves. The greater internal area of the thicker sun leaves makes a substantial contribution to their greater photosynthetic capacity. These and other differences between sun and shade leaves are listed in Table 6.3. Shade leaves show features which allow them

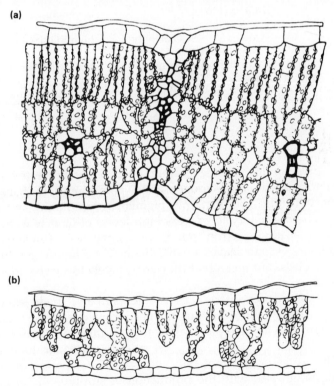

Figure 6.16 Cross-sections of sun and shade leaves of sugar maple (*Acer saccharum*): (a) leaf from the sunny south side of a tree (note the thick cuticle, long palisade cells and number of cell layers); (b) leaf from the centre of a canopy.

Table 6.3 General characteristics of sun and shade leaves.

Descriptive level	Shade characters	Sun characters
morphological	often, a mosaic leaf arrangement, phototropic leaf petioles	
anatomical	fewer stomata, leaves broader, thinner: fewer palisade layers, shorter palisade cells, more intercellular spaces (low $A_{mes} : A$);	more stomata, leaves thicker: more palisade layers, longer cells, tighter packing (high $A_{mes} : A$);
	thin cuticle	thick cuticle, often hairy, often pigmented
ultrastructural	larger chloroplasts, more grana, phototactic chloroplasts.	smaller chloroplasts, fewer and smaller grana
biochemical	more chlorophyll, more chlorophyll b, less redox coenzymes, lower RUBISCO activity, low rates of respiration, low light compensation point, low light saturation levels	lower chl $a : b$ ratio, more redox coenzymes, high RUBISCO activity, high light compensation point, high light saturation levels

both to survive in low light levels (e.g. a low respiration rate and thus a low photosynthetic light-compensation point), and to harvest low levels of light (e.g. leaf anatomy, pigment composition), but they do not have the machinery to process high levels of light (e.g. they have low levels of the electron transport coenzymes). Conversely, the characteristics of sun leaves enable them to be highly productive in high light levels, but mean that they are unlikely to survive in low light conditions.

These light-induced differences between sun and shade leaves show a better correlation with the total fluence received over the whole light period than with any simple peak of fluence rate: the greater the total daily fluence, the greater the $A_{mes} : A$ ratio. In particular, high levels of blue light are associated not only with the development of sun-type chloroplasts but also with the production of sun-type or xeromorphic leaves.

Stomata, a major structural and functional feature of leaves, are discussed in Box 6.3.

Box 6.3   Stomata

Stomata are of obvious importance to the processes of photosynthesis and transpiration. However, in a wider sense, they are also important as a means by which the internal environment of the plant can be regulated and adjusted in relation to a continuously changing external environment (Zeiger 1983). For example, opening the stomata has significant cooling effects; closing them, minimizes the entry of pollutants and pathogens.

Stomata were an early development in the land flora: they are present on the sporophytes of the Pteridophyta and Bryophyta, as well as on higher plants. Their occurrence on the leaf surface is not only of obvious advantage for the uptake of CO_2, but, less obviously, this location is also the best point at which to exert control over the water balance of the plant. Since this interface contains the greatest single drop in water potential (ΨH_2O) in the whole soil–plant–air continuum, it is the most advantageous point at which to control the movement of water from the soil to the atmosphere (fall in ΨH_2O : soil $\rightarrow$ leaf = 30 bars; leaf $\rightarrow$ air = 90 bars).

Stomatal opening arises from the increased turgor of the guard cells in relation to the other cells of the epidermis. Potassium is accumulated in the vacuole of the guard cells, charge balance being maintained by concomitant chloride uptake and malate synthesis. This results in a decreased ΨH_2O in the guard cells and a consequent inflow of water. The potassium accumulation is thought to occur in response to the action of a proton pump on the cell membrane: the uptake of K^+ is coupled to an ATP-driven efflux of H^+.

Four factors influence stomatal aperture: an endogenous rhythm to opening and closing, the water balance of the plant, CO_2 concentration and light. The general operation of circadian rhythms is described in Chapter 8. Water stress results in stomatal closure possibly through effects of increased concentrations of abscisic acid on K^+ uptake or on the ATPase-proton pump. The mechanism by which increased levels of CO_2 bring about closure is unknown. Guard cells do not fix CO_2 via the Calvin cycle, but do incorporate it into malate through the action of PEP carboxylase. However malate synthesis is too slow to account for the rapid effects of CO_2 on stomatal aperture. Possibly CO_2 affects some aspect of membrane permeability. The action of CO_2 on stomatal closure can generally be overridden by the more dominant effects of light on opening.

Light induction of stomatal opening has action maxima in the blue and red regions of the spectrum, and for many years was attributed to the photosynthetic activity of chlorophyll. However, it is now apparent that light affects stomatal aperture through the actions of more than one photosystem. Involvement of a distinct photoreceptor in the blue region is suggested by several features:

(1) Relatively greater photon effectiveness of blue light on stomatal aperture (including sensitivity to much lower fluence rates of blue light than of red light).

135

(2) Responsiveness to pulses of blue light, but not to pulses of red light.
(3) Swelling of isolated guard cell protoplasts in blue light, but not in red light.
(4) Lack of effect of the photosynthetic inhibitor, DCMU, on blue light-induced opening, but inhibition of red light-induced opening.

Therefore, it is thought that red light causes opening through the photosynthetic production of ATP and $NADPH_2$, which in turn have effects on the proton pump and on malate synthesis. Blue light itself does not seem to act as an energy supply. Its effects are inhibited by KCN or anaerobic conditions, i.e. its action depends on some other (oxidative) supply of energy. Possibly, blue light has direct effects on the membrane ATPase-proton pump.

There is a strong possibility that phytochrome is also involved in the regulation of stomatal aperture. The guard cells of the orchid *Paphiopedilum* contain no chloroplasts, but still respond positively to red and blue light. Further, in many systems far-red irradiation prevents stomatal opening.

It would seem advantageous to have more than one photosystem regulating stomatal aperture. If photoregulation were wholly through the photosynthetic system, then it would probably oscillate in relation to light intensity (e.g. in response to cloud effects). The action of another system possibly smooths out such effects. The greater sensitivity to blue light also ensures that the stomata are open at dawn (in response to the relative blue light enrichment) in readiness for the initiation of photosynthesis (Zeiger 1983).

6.6 SUMMARY

(1) Photomorphogenesis encompasses a multiplicity of light responses which are highly specific for a particular organ at a particular time. These patterns of spatial and temporal specificity are determined by the specific states of competence of the responding cells, rather than by specific properties of light or photoreceptor.

(2) At the level of the seed, the light quality of the environment is an important factor both in the induction of dormancy and the stimulation of germination. Germination is promoted by P_{fr}; seed light responses are ecologically significant as mechanisms which help to place seedlings spatially and temporally in environments suitable for further growth and development.

(3) At the organ level, light causes both stimulation and inhibition of growth; this possibly results from a P_{fr}-mediated acceleration of the normal sequence of cell development. The action of light

on cell growth possibly involves effects on cell wall properties.

(4) The light-growth responses of the grass seedling depend on the type of organ, its age and the type of light treatment. Brief exposures to light stimulate growth in young coleoptiles and inhibit growth in older coleoptiles; the mesocotyl is much more sensitive to light and is inhibited at very low fluences.

(5) The opening of the plumular hook in dicot seedlings is controlled by light. Photocontrol of seedling growth is mediated through both phytochrome and the more rapid effects of a blue-light receptor; generally, light acts to inhibit stem growth in seedlings.

(6) The light growth responses of green' plants are quantitatively and qualitatively different from those of etiolated seedlings; green seedlings are generally more sensitive to fluence rate, but have lost sensitivity to continuous far-red irradiation; growth responses to changes in the light environment are particularly evident in shade-avoiding species, where both decreased R : Fr ratios and decreased fluence rates give increased growth rates.

(7) The greening process comprises many developmental changes at many different levels of organization; chloroplast development involves both the induction of pigment biosynthesis and ultrastructural reorganization and synthesis; in all of these light-induced effects, protochlorophyllide plays a key regulatory role; phytochrome and a blue-light receptor are also involved.

(8) Light has a marked influence on leaf size and shape, through stimulatory effects on both cell division and expansion. Light quantity, in particular, influences leaf area, thickness, cell number and cell size. Large differences in form and functioning are seen between sun and shade leaves.

(9) Light affects stomatal opening through more than one photo-system. Red and blue light act through the light reactions of photosynthesis; another large proportion of the effects of blue light are due to the action of a specific blue light receptor. Phytochrome is also possibly involved.

FURTHER READING

Shropshire, W. & H. Mohr (eds.) 1983. *Photomorphogenesis*. Encycl. Plant Physiol. NS 16 A & B. Berlin: Springer.

Smith, H. (ed.) 1981. *Plants and the daylight spectrum*. London: Academic Press.

CHAPTER SEVEN

Orientation in space: phototropism

7.1 INTRODUCTORY COMMENTS

Plants are in the main non-motile organisms. Moreover, the raw materials necessary for their growth and development are present in the environment in relatively dilute amounts. Therefore, to maximize their uptake of nutrient and energy, plants must continually monitor and adjust the orientations of their various organs. One way that this is done is through **tropisms** – directional responses to environmental stimuli. The majority of tropic responses occur through differential growth, although heliotropism (Sun-tracking, Ch. 1) often involves reversible turgor changes (Smith 1984). The distinguishing feature of a tropic movement, however, is that the direction of response bears some relation to the direction of the stimulus, rather than being determined by the anatomy of the responding organ. The actual direction of the response can be towards (*positive*) or away from (*negative*) the stimulus; or it can be at a right angle (*dia*-tropic) or at some other angle (*plagio*-tropic) to the direction of stimulus.

The major environmental factors involved in the tropic orientation of plant organs are light and gravity. In certain situations, other factors can play significant rôles – for example, touch (thigmotropism), as in the tendrils of some climbing plants; and chemicals (chemotropism), as during growth of the pollen tube down the style. (In older texts, hydrotropism, electropism and magnetropism are also described.)

In phototropism, the differential growth occurs in response to a non-homogeneous distribution of light, rather than strictly to the direction of light. The response is particularly evident in rapidly growing material, like seedlings, and in leaves and reproductive organs, including the sporophores of mosses and many fungal sporangiophores. In general, stems and aerial organs are positively phototropic, roots and other subterranean organs often show negative responses, and leaves are mainly plagiophototropic. However, there are very many exceptions to these generalities. The tendrils of many climbing plants are

138

negatively phototropic. In *Hedera helix* (ivy), the stem is negatively phototropic, although the leaf petioles show positive responses (Fig. 1.3). The tropic behaviour of some organs even changes with age or stage of development; for example, the flower stalk of *Cymbalaria muralis* (ivy-leaved toadflax) becomes negatively phototropic after fertilization (to result in the burial of the seed pod in a dark crevice of the walls on which this plant is usually found).

Despite this diversity of response, phototropism in higher plants has really only been intensively investigated in a limited range of higher plant material, namely the grass coleoptile, particularly that of oats – although see Box 7.1. (Since the coleoptile encloses and protects the true leaves of the seedling during upward penetration through the soil, its possession of an extremely sensitive light guidance system is a decided advantage.) The types of light treatment used in phototropic experiments have also been limited, being restricted in the main to brief exposures at very low fluence rates, very different from natural conditions. Therefore, it may be that we know little about the phototropic responses of some more permanent, de-etiolated organs in the natural light environment.

7.2 SIGNAL PERCEPTION: LIGHT RESPONSES[1]

7.2.1 *Region of perception*

A traditional tenet of tropic behaviour is that there is some distinction between the perception of the stimulus and the consequent growth response. Darwin concluded that there was an actual spatial distinction between regions involved in perception and response, from his experiments with coleoptiles: he noted that in seedlings which were buried to different depths in fine sand, curvature still occurred in the more basal regions of the coleoptile even when just the tip of the organ was exposed to unilateral light. Subsequent work has confirmed that the apical regions of the coleoptile show the greatest phototropic sensitivity (although it is not the extreme apex which is most sensitive, but a region about 100 µm behind the tip). However, and more significantly, *sensitivity is not restricted to these apical regions*: there is a decreasing gradient of phototropic sensitivity throughout the whole length of the organ.

Temporal distinction between the processes of phototropic perception and response can be obtained through manipulation of temperature: if coleoptiles are exposed to a unilateral light stimulus at 2°C, curvature can be subsequently expressed in darkness at 20°C. In fact, after

[1] See Darwin (1880), Dennison (1979, 1984).

Box 7.1 Phototropic behaviour of the *Phycomyces* sporangiophore

Another organ whose phototropic behaviour has been intensively studied is the sporangiophore of the fungus, *Phycomyces blakesleeanus* (Dennison 1979). This reproductive organ is a fast-growing unicellular structure about 1 cm long. It shows a positive response to unidirectional light within a few minutes (Fig. 7.1), behaviour of obvious significance to the efficient dispersal of its airborne spores. Indeed, it has recently been shown that the sporangiophore of this fungus is one of the most photosensitive biological organs (Iino & Schäfer 1984). It responds to a 10-second light-pulse which delivers a total fluence of only 10^{-7} J m^{-2}; this represents a sensitivity to blue light more than 1000 times greater than that of the grass coleoptile.

With major peaks at λ 365 nm and λ 455 nm, the action spectrum for the phototropic response in *Phycomyces* shows a general similarity to those of other blue-light induced responses (Presti and Delbruck 1978). However, recent analyses have shown that there are other minor peaks of activity (at λs 414 nm, 491 nm and 650 nm) and that the shapes of the fluence-response curves are wavelength-dependent (Galland 1983). These features strongly suggest that more than one photoreceptor is involved in the directional light responses of the sporangiophore.

It is generally accepted that this phototropic mechanism involves a phenomenon called the *light growth reaction* (Dennison 1979). That is, exposure of the sporangiophore to light, or even to a sudden increase in fluence rate, brings about a (temporary) *increase* in growth rate; therefore, an asymmetry in fluence rate between the two sides of the organ gives an asymmetry in growth rate across the organ. Curvature is *towards* the light source, due to the establishment of a higher fluence rate (and therefore higher growth rate) on the side *farthest* from the light source. The light is, in fact, focused onto the far wall by the refractive properties of the sporangiophore itself. This was concluded from observations that the phototropic behaviour was changed in situations that would tend to negate the lens effect of the biological material (see below, Fig. 7.3):

(1) Immersion of the sporangiophore in paraffin oil which has a refractive index greater than that of the organ, removes the focusing effect. In this situation, the phototropic response becomes negative.

(2) Unilateral irradiation of the organ with light which is first passed through an external lens weakens the focusing effect of the biological material and also results in negative curvature.

(3) The phototropic response to UV irradiation is itself negative. This is due to the fact that the sporangiophore contains a high content of gallic acid, which strongly absorbs UV radiation. The 'light' gradient across the organ in this case, therefore, results from absorption, rather than from refraction. Intensity, and thus growth rate, is higher on the side nearer the UV source. A similar effect can be obtained in blue light if the sporangiophore is infiltrated with a blue-absorbing dye.

Figure 7.1 Phototropic curvature of the sporangiophore of *Phycomyces*. The source of blue light is on the left, photos taken at 2 min. intervals.

phototropic stimulation, the material can be stored in darkness at 2°C for at least seven hours and still retain the ability to develop subsequent curvature at 20°C. (Not all species, however, exhibit this phototropic 'memory' effect.)

7.2.2   Action spectrum

There is an apocryphal report that in the 19th century a Frenchman observed that plants growing behind bottles of red wine did not bend towards the light; and Darwin noted that light transmitted through a solution of potassium dichromate did not cause any 'heliotropic curvature'. However, it was A. H. Blaauw, in 1909, who was one of the first specifically to demonstrate that blue light is the most effective for induction of a phototropic response. Subsequent determinations of action spectra by various investigators have resulted in agreed action maxima at λ 475 nm, λ 450 nm, a shoulder at λ 420 nm and a peak in the UV at λ 370 nm (Fig. 7.2a). Phototropic curvature is also induced by unilateral irradiation with UV at λ 280–295 nm (Fig. 7.2b), but it is not yet clear whether this is part of the same overall response.

7.2.3   Photoreceptor

As with other responses to blue light, the molecular nature of the photoreceptor is unresolved. Initially, in the 1930s, some form of *carotenoid* was thought to be the phototropic receptor, on the basis of circumstantial evidence such as:

(1) the ubiquity of carotenoids in biological materials;
(2) their relationship to the retinal pigments of vision;
(3) the spectral match between phototropic action and carotenoid absorption;
(4) the localized concentration of carotenoid in the phototropically most sensitive sub-apical region of the coleoptile.

Further evidence that carotenoids can act as photoreceptors, at least in

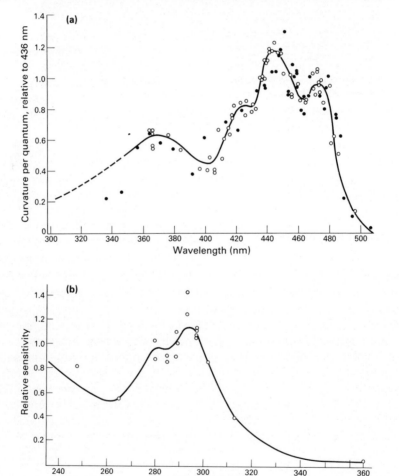

Figure 7.2 Action spectra for phototropic curvature in *Avena* coleoptiles. (*a*) First positive response (*b*) Base curvature in response to unilateral UV irradiation.

some plants, is provided by the recent demonstration (Foster *et al.* 1984) that a rhodopsin is the receptor for phototaxis in *Chlamydomonas*: when chemical analogues of the retinal chromophore were supplied to 'blind' mutants, not only was normal photobehaviour restored, but the consequent action spectra were consistent with the particular analogues supplied.

General opinion at present, however, favours some kind of *flavin* as phototropic receptor, largely on the basis of:

(1) the spectral match between phototropic action and flavin absorption in the UV region;
(2) the retention of phototropic sensitivity in *Phycomyces* mutants which have carotenoid synthesis blocked at six different loci;
(3) loss of phototropic sensitivity in tissues treated with chemicals such as potassium iodide and phenylacetic acid, which quench flavin, but not carotenoid, photoexcitation; the gravitropic responses of such treated tissues are largely unaffected, suggesting that the effects on phototropism are not simply due to chemical effects on growth (Vierstra & Poff 1981a).

(However, a recent investigation examined the phototropic responses of maize coleoptiles which had been treated with the herbicide Norflurazon, an inhibitor of carotenoid synthesis (Vierstra & Poff 1981b). The results were not unequivocal: the phototropic response to λ 380 nm was unaffected, while phototropic sensitivity to λ 450 nm was reduced. This was taken to indicate that, while 'bulk' carotenoid is not the actual phototropic receptor, carotenoids are involved in the response to λ 450 nm, by acting as a screen to increase the light gradient across the tissues.)

7.2.4 Light gradient

For a phototropic response to occur, the light distribution around an organ must be perceived as a light gradient across the tissues. Non-homogeneous perception of light can be accomplished in various ways in biological materials. In the first place, if the photoreceptor is fixed in a **dichroic orientation**, the perception of light within the cell will be non-uniform (Haupt 1983). This is illustrated in the diagrams of Figure 7.3a (these diagrams are wholly derived from the elegant work of Haupt on phytochrome-mediated responses to polarised light in the alga, *Mougeotia*, see Ch. 5). In Figure 7.3a, the transition moments of the photoreceptor molecules are broken into longitudinal and perpendicular surface-parallel components, and those which can absorb light

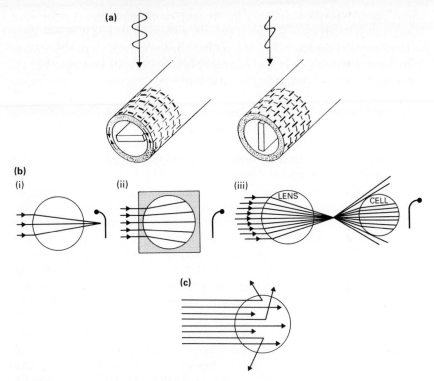

Figure 7.3 Non-homogeneous light perception in biological materials (a) Dichroically-oriented photoreceptor, e.g. phytochrome in a *Mougeotia* cell. The transition moments of the receptor are here decomposed into longitudinal and perpendicular surface-parallel components; those which can absorb are shown in bold dashes and the P_{fr} distributions resulting from irradiation with red light plane polarised in two different orientations are shown by dots. (b) Light refraction, e.g. sporangiophore of *Phycomyces*. Behaviour of light: (i) in a sporangiophore in air; (ii) in a sporangiophore submerged in a medium of greater refractive index; (iii) when the light is first passed through another focusing lens. (c) Light attenuation by absorption and scattering.

under the particular conditions illustrated are shown in bold dashes; the resulting distributions of photo-transformed receptor are shown by dots in the cross sections. If the cell is irradiated with polarized red light, then, due to differential (dichroic) light absorption, particular distributions of P_{fr} will arise – in the two extreme situations illustrated here, a uniform and a non-uniform distribution. Hence, in normal unpolarized light (a 'mixture' of situations (i) and (ii) in Figure 7.3a), the distribution of P_{fr} must also be non-uniform. Note, however, that dichroism in itself is unlikely to result in light being perceived as a gradient across a cylindrical cell or organ; effects are similar at the front and rear.

An actual light gradient across an organ can be created or intensified by just two basic types of effect (Poff 1983):

(1) **Light refraction** (Fig. 73b), in which the change in refractive index between an organ and its surroundings can result in a difference in light intensity between the two sides of the organ; in the sporangiophore of *Phycomyces*, the light is literally focused onto the wall farthest from the light source.

(2) **light attenuation** (Fig. 7.3c) which itself results from two types of process: *screening*, through light absorption by pigments, both by the photoreceptor itself and by other inactive pigments; and *scattering*, through reflection (by scattering particles larger than the light wavelengths) and diffraction (by particles smaller than the light wavelengths).

There is no evidence for any refractive 'lens effect' in unilaterally irradiated tissues of higher plants. On the contrary, it has recently been shown that substantial light gradients exist across unilaterally irradiated coleoptiles and hypocotyls. (These measurements have been made both by inserting microfibre optic probes into the tissues, and by measuring light transmission through tissue slices of different thicknesses (Cosgrove 1985 and references therein). The sizes of these gradients vary according to the method used to measure them, from 5 : 1 to 50 : 1. It is generally felt that gradients of these magnitudes are unlikely to arise solely from screening effects, and that therefore light scattering also plays a major role. The degree of scattering is inversely related to the wavelength of light – that is, scattering effects impose a steeper gradient on blue light than on red light. However, the involvement of the blue region as the directional light signal is probably due to more than just this effect.

Besides perception of a light gradient, a differential response across a tissue also requires that there be a 'transduction gradient': the effects of light must remain highly localized. It is this aspect that may account for the lack of directional response to red light by most organs in higher plants. Even though a gradient of red light can occur across a hypocotyl, and be measurably manifest in terms of anthocyanin formation (Mohr 1972, p. 208), a gradient of growth is not generally obtained. It may be that phytochrome initiates some sort of transmissible effect on growth which is 'averaged out' over the tissue, while effects of blue light remain more localized.

Unilateral red light can, however, have a phototropic effect on the mesocotyls of maize seedlings (Iino *et al.* 1984); this is an example of a phytochrome-mediated 'Very Low Fluence' response, where presumably the extreme sensitivity of the system somehow enables it to respond to a gradient of P_{fr}. Red light can also induce directional growth responses in some lower organisms, for example, in moss protonemal filaments (Mohr, 1972, p. 206). But in this case, the mode of response is quite different: red light induces a change in position of the growing point in a single apical cell, possibly through differential light perception within that cell by dichroically oriented phytochrome molecules.

7.2.5 Light treatment and fluence-response

The extent of phototropic curvature is a function of many factors, including the species, its age and growth rate, and, not least, the type of light treatment. Under continuous unilateral light, oat coleoptiles show some curvature within 15 minutes; by 100 minutes, the extent of bending is such that gravitropism becomes a serious antagonistic effect. Darwin used continuous exposures to natural light and found no effect of intensity on the extent of response. A. H. Blaauw (cited in Dennison 1984) initiated a more quantitative approach; by the use of *limited exposures to low levels of light*, he related the extent of response to the amount of light received. At light levels which give a threshold response, he found reciprocity between fluence rate and time over an extremely wide range of times (indicating the involvement of only one photoreceptor in the response). However, subsequent investigation of the phototropic fluence response curve *at higher fluence rates* revealed a much more complex situation, with regions of maximum, minimum and even negative responses (Fig. 7.4). For example (Fig. 7.4, curve 3), at a fluence rate of 0.038 W m^{-2} (i.e. 0.038 J s^{-1} m^{-2}), if the fluence (dose) is increased by extending the duration of exposure to light, the response increases up to a maximum at around 0.1 J m^{-2}; further increase in fluence gives a decline and even curvature away from the light source (negative response) before the extent of positive curvature again increases. These different types of response are termed, respectively, *first positive*, *first negative* and *first positive* curvatures. However, at lower fluence rates (Fig. 7.4, curves 1 and 2), longer exposures are necessary to achieve the equivalent fluences; under these conditions distinction between first and second positive type curvatures becomes blurred, and the negative response is lost. Reciprocity between time and fluence rate is only shown for first positive and first negative curvatures. Under conditions of second positive curvature, i.e. under 'long' exposures (at a fluence rate of 0.03 W m^{-2}, anything longer than four minutes), there is no reciprocity, and *response is related only to the length of exposure*. (Phototropic responses under natural conditions are, of course, due to second positive type curvatures, but it is the first positive type which has been most intensively investigated.)

From analyses of this complex fluence-response curve, Zimmerman & Briggs (see caption to Fig. 7.4) suggested that the different responses may derive from the actions of more than one light-driven reaction.

However, the action spectra for first and second positive types of curvature are similar, and the forms of the fluence response curves are similar at the different action maxima of λ 370 and λ 450 nm. Therefore, at present, the general consensus is that only one photoreceptor is involved.

Further investigation has revealed even greater complexity in fluence-response. Two Dutch investigators, O. H. Blaauw and G. Blaauw-Jansen (1970), measured coleoptile curvature as a function of a great many fluence rates and at many exposure times. The three-dimensional depiction of their results is referred to as the *fluence-response surface* (Fig. 7.5); from this, the response can be considered in relation to fluence rate (increasing rear to front), or in relation to length of exposure (increasing left to right). At the highest fluence rates (front, 10 Wm^{-2}), increasing periods of exposure result in the familiar patterns of regions of maximum and minimum responses (maxima are referred to by these authors as regions A, B and C). At lower fluence rates (rear), regions A and B shift to longer exposures (i.e. reciprocity holds), but region C does not; thus, region C is analogous to the original second positive type of curvature. When the response is considered in relation to period of exposure, short exposures (left side) give good curvatures at high fluence rates; but at longer exposures (right side) high fluence rates give less curvature than low fluence rates. (This could indicate that at high fluence rates, a time-dependent inhibitory effect is being superimposed on the response.)

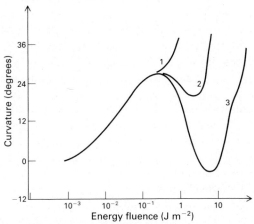

Figure 7.4 Fluence response curves for phototropism in coleoptiles. Curves 1, 2 and 3 were determined at fluence rates of, respectively, 3.84×10^{-4}, $\times 10^{-3}$ and $\times 10^{-2}$ W m^{-2} at λ 436 nm.

7.2.6   Tip and base responses

Obviously, these complexities of fluence response are to a large extent man-made, inasmuch as they derive from the mode of experimentation and type of light treatment. Besides these types of response, which are distinguishable by type of light treatment, there may also be different responses distinguishable in different regions of the coleoptile. These were originally characterized by Thimann and Curry as tip and base responses (Fig. 7.6). The 'tip' response refers to a type of curvature induced by low fluence (in the region of first positive) and slowly propagated along the coleoptile as far as the mid-point of the organ. The 'base' response describes a faster curvature at high fluences (in the region of second positive), which appears along the whole length of the coleoptile. (It is also a base type of response that is induced by unilateral irradiation with UV at λ 290 nm.)

It has since been suggested by Blaauw & Blaauw-Jansen (1970), that such categorization into distinct types of tip and base response is spurious, and that the differences are simply due to effects of red light on the phototropic system. Pretreatment of the tissue with red light decreases the sensitivity of the first positive response (shifts it to higher fluences), but increases the sensitivity of the second positive response (shifts it towards lower fluences). Uncontrolled exposures to red light may therefore account for the differences in sensitivities and responding regions in the tip and base responses.

7.2.7   Dicotyledons

The phototropic responses of dicot seedlings have not received nearly as much investigation. Light-directed orientation in a dicot may involve a more complex situation, since more than one system may be involved. Red light can itself induce directional growth in, for example, cucumber seedlings in which one cotyledon is artificially darkened (Fig. 7.7). This effect seems to be due to the transmission of some growth inhibition effect from the irradiated cotyledon to the hypocotyl tissue directly below it (see also Schwartz & Koller 1980). It may be that, in nature, the red light fluences received by individual leaves have some influence on overall orientation, through control of growth in particular regions of the stem. In contrast, the effects of blue light on phototropism in dicot seedlings involve light perception by the stem or hypocotyl tissues themselves.

From the limited number of blue light fluence-response studies that have been carried out, it seems that dicot seedlings are much less sensitive than coleoptiles to unilateral irradiation. In general, dicots only begin to show curvature after light exposures of around

Figure 7.5 The fluence response surface for phototropism in *Avena* coleoptiles. Curvature is shown as a function of both fluence rate and exposure time; from back to front are increasing fluence rates and from left to right are increasing exposure times.

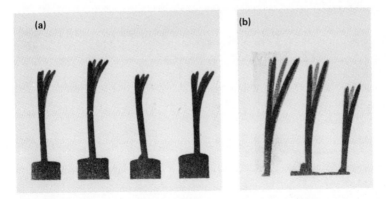

Figure 7.6 The two types of phototropic response in *Avena* coleoptiles: (*a*) tip response; (*b*) base response.

149

60 minutes, although there are exceptions such as *Fagopyrum* (buck-wheat) which respond much more rapidly (Ellis 1984). Significant phototropic curvature in dicot seedlings is due to a type of second positive response; any response analogous to first positive curvature is either not present or only very weakly expressed (Fig. 7.8). In many species (e.g. cress, lettuce, radish), etiolated seedlings are even less phototropically responsive than corresponding green seedlings (Fig. 7.8). Intriguingly, the forms of curvature shown by these two

Figure 7.7 Photocurvature in cucumber seedlings in non-unilateral light. Curvature was induced over a period of 8 hrs by covering the left-hand cotyledons with aluminium foil.

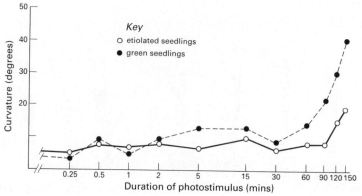

Figure 7.8 Phototropic curvature in cress seedlings in response to increasing exposure to unilateral light. Seedlings were exposed to 2 μmol m^{-2} s^{-1} blue light for indicated periods.

types of seedling, etiolated and green, are seen to be quite distinct when examined by time-lapse photography (Fig. 7.9): the more slowly developing curvature of etiolated seedlings starts at the tip and progresses down the hypocotyl, while the more rapid curvature of green seedlings initially appears along the length of the hypocotyl. These different forms of response seem analogous to the tip and base responses of the coleoptile.

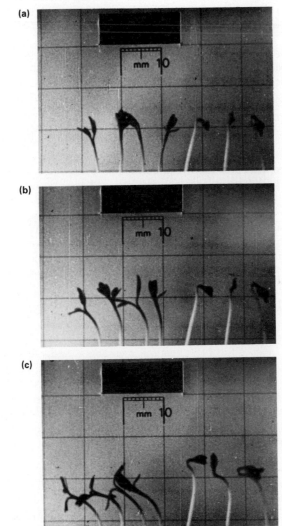

Figure 7.9 Time-lapse sequence of phototropic curvature in hypocotyls of etiolated (3 plants on right) and green (4 plants on left) cress seedlings. Plants were exposed to 3 μmol m^{-2}s^{-1} continuous, unilateral blue light. (a) 0 time (darkness); (b) 35 min. light exposure; (C) 80 min. light exposure.

Box 7.2   Historical aspects of phototropic research

In many ways, the historical aspects of research into phototropism are as interesting as the process itself. Two investigators made important contributions to the early characterization of the phenomenon – it could be said that while Darwin bequeathed the grass coleoptile to phototropic resarch, A. H. Blaauw established the quantified light treatment. However, more significant than their technical contributions are the different emphases, or ways of thinking about the process, that their approaches engendered.

On the basis of his observations of distinction between region of light perception and region of growth response, Darwin, in 1880, suggested that some sort of 'transmissible influence' controlled growth. On the other hand, Blaauw, from his equally classic investigations which related the extent of response to the level of light stimulation, emphasized a direct effect of light on growth and, in 1919, proposed that asymmetry in light reception directly gave an asymmetry in growth rate. (Such a situation is still thought to account for photocurvature of the *Phycomyces* sporangiophore.) In subsequent work on higher plants, of course, it was Darwin's *chemical messenger* concept that received the greatest development, culminating in the discovery of auxin and the Cholodny–Went hypothesis. (Although present concepts regarding the regulation of plant growth are generally related to the Darwin view, the ideas of Blaauw may be more appropriate to the rapid effects of blue light on growth.)

The Cholodny–Went hypothesis to account for tropic curvature developed through a series of what have come to be viewed as classic experiments.

(1) *Boysen-Jensen* (*circa* 1913) studied the extent of curvature in coleoptiles which had mica diffusion-barriers inserted in various orientations (Fig. 7.10). These experiments supported the possible existence of a diffusible chemical messenger and, furthermore, emphasized the importance of the shaded side of the unilaterally irradiated organ.

(2) *Paal* (*circa* 1919) showed that, while removal of the coleoptile tip greatly lessened the extent of tropic response, its asymmetric replacement resulted in curvature away from that side on which the tip had been replaced (Fig. 7.10). Such experiments supported the idea that the tip might be the site of production of a diffusible growth substance, and further, that tropic curvature might arise from asymmetric distribution of this chemical.

(3) *Cholodny* and *Went* actually worked independently, in the 1920s. Cholodny, from his studies initially on root gravitropism and subsequently on phototropism in coleoptiles, postulated that tropic curvature derived from lateral movement of growth promoter in the tip of the responding organ. Went, with his newly developed bioassay techniques, showed that a diffusible growth promoter,

152

auxin, could be obtained from plant tissues, and that there was about twice as much of it on the shaded side of a phototropically curving coleoptile.

All these experiments not only led to a hypothesis to account for tropic curvatures, but the discovery of auxin had, and continues to have, a tremendous influence on the theory and practice of plant growth regulation. Indeed, a direct line of descent can be traced between the initial (esoteric?) experiments of Darwin on a plant's directional responses in a darkened room, and the present use of many agrochemicals for the commercial manipulation and control of plant growth.

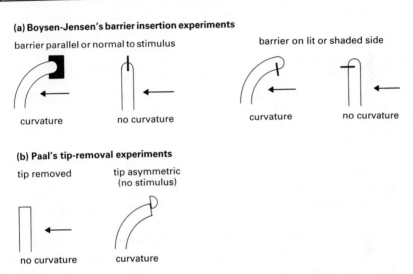

(a) Boysen-Jensen's barrier insertion experiments

barrier parallel or normal to stimulus

barrier on lit or shaded side

curvature   no curvature   curvature   no curvature

(b) Paal's tip-removal experiments

tip removed   tip asymmetric (no stimulus)

no curvature   curvature

Figure 7.10 Some early experiments on phototropism in coleoptiles; arrows indicate the direction of irradiation.

Some of the historical aspects of the phototropic responses of plants are discussed in Box 7.2.

7.3 SIGNAL PROCESSING: GROWTH RESPONSES[2]

7.3.1 *Hormonal involvement*

For over 50 years, the Cholodny–Went hypothesis has been generally accepted as the explanation of the means by which a directional

[2] See Dennison (1979, 1984), Firn & Digby (1980).

environmental stimulus produces a directional growth response. In its original form, the hypothesis states that:

(1) tropic stimulation causes *lateral migration of auxin in the tip* of the responding organ;
(2) the asymmetry in auxin distribution is *transported longitudinally* along the organ to affect growth in other regions of the organ;
(3) in shoots, curvature derives from *accelerated growth* on the convex side of the curving organ.

The asymmetry in auxin distribution is thought to result from auxin movement, rather than from effects on its synthesis or destruction, since the total amounts of auxin remain similar before and after tropic stimulation. The tip of the coleoptile is implicated as the region where such movement occurs, not only from the original suggestions of Cholodny, but also by the results of auxin diffusion experiments (Fig. 7.11). When coleoptile tips were placed on blocks of agar and subjected to unilateral irradiation, different amounts of auxin diffused out of the lit and shaded sides; this auxin differential in the receiving blocks was not obtained if an impervious barrier was inserted in the tip to prevent lateral movement of auxin. Asymmetries in auxin distribution after tropic stimulation, have been found in the actual tissues themselves on a number of occasions. These cases include not only determinations of endogenous auxin but also demonstrations of subsequent gradients in exogenously supplied radiolabelled auxin (Table 7.1). Generally such gradients consist of ratios in auxin content of the order of 35 : 65 between the two sides of a curving organ. (In phototropically stimulated sunflower supplied with radiolabelled

Table 7.1 Summary of the asymmetric distribution of ^{14}C–IAA in maize coleoptiles after unilateral light treatment (from Thimann 1977).

Light dosage	Plant material		Amount of IAA transported normal	Asymmetry found		
	Illuminated	Analysed		No. of expts	Lighted side	Shaded side
First positive	tips	agar receivers	*ca.* 0.1 ×	4	24	76
First positive	tips	agar receivers	0.15–0.4 ×	4	25	75
First positive	tips	agar receivers	0.5–1.2 ×	5	34	66
First positive	tips	tissue halves	0.1–0.25 ×	8	35	65
Second positive	tips	agar receivers	*ca.* 0.5 ×	20	34	66
Second positive	subapical section	agar receivers	1.4 ×	5	46.1	53.9
Second positive	subapical section	agar receivers	0.5 ×	7	46.9	53.1
Second positive	subapical section	agar receivers	0.1 ×	6	46.0	54.0
Second positive	subapical section	agar receivers	0.02	12	41.8	58.2

gibberellin rather than auxin, a ration in radiolabel of 12 : 88 was found between the lit and shaded sides, respectively.)

The mechanisms by which such lateral migration of growth substance is initially brought about by unilateral irradiation (or by gravitropic stimulation), and how the asymmetry is maintained along the length of the organ, are not known. There are differences in electrical polarity across a phototropically stimulated coleoptile, but these are generally thought to be a result of, rather than a cause of, the auxin asymmetry.

Besides this large gap in knowledge of the transduction sequence between light perception and auxin movement, there are also problems with many of the original precepts of the Cholodny–Went hypothesis (discussed most notably by Richard Firn and John Digby (1980) of York University):

(1) *The coleoptile tip.* Decapitated coleoptiles curve in response to phototropic stimulation; similarly, coleoptiles with the apices capped to exclude light also respond to unilateral irradiation. (In both cases, the effective light exposures are in the region of second positive curvature.)

(2) *Stimulus transmission.* The York investigators report that all regions of a tropically stimulated organ respond simultaneously,

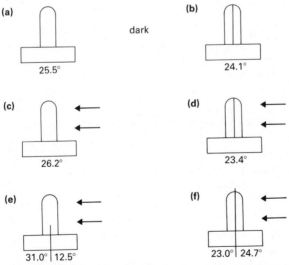

Figure 7.11 Auxin diffusion into agar blocks from variously treated coleoptile tips of maize. The number under each block indicates the degrees of curvature produced by that block in the Went curvature test for auxin; the vertical lines through some of the tips represent impermeable barriers; the arrows indicate the direction of irradiation. (Twice as much auxin was obtained in (e) and (f) because each block had been in contact with 6 half-tips, the equivalent of the 3 whole tips of (a)–(d).

and that the region of curvature does not move along the organ. (This is strongly contested by many other workers.)

(3) *The auxin gradient.* Tropic curvature has occurred in many cases where an auxin gradient could not be demonstrated; furthermore, it has been questioned whether the gradients that have been observed, are either of the appropriate magnitude, or occur soon enough, to account for the growth differentials that give rise to curvature.

(4) *The growth responses.* The actual changes in growth rate occur very rapidly, in some cases within as little as five minutes from the onset of tropic stimulation. They also universally involve, besides growth acceleration, phases of strong growth inhibition.

Whether the extent of these difficulties means that the hypothesis must be totally discarded remains to be seen. But certainly, the rapid changes in growth rate, inhibitory as well as stimulatory, serve as a reminder of the original ideas of A. H. Blaauw, namely, that phototropism resulted from direct effects of light on growth.

7.3.2 *Growth responses*

Blaauw investigated growth in both the fungal sporangiophore and the coleoptile, and in each type of organ he noted responses that he termed 'light growth reactions'. The transient increase in growth rate of the *Phycomyces* sporangiophore in response to light, and the light-focusing effects of the sporangiophore itself, which are responsible for its curvature towards the light source, have already been described (Box 7.1). In the oat coleoptile, however, Blaauw observed a temporary *decrease* in growth upon exposure to omnilateral light, followed by a transient growth stimulation. Again he reasoned that, under unilateral irradiation, differential light reception between the two sides of the organ would result in a differential growth rate across the organ. But in this situation, the side *nearer* the light source must receive the greater fluence rate; thus its *decreased* growth rate would result in curvature of the coleoptile towards the light source.

It seems highly likely that these light growth reactions described by Blaauw are related to the rapid growth responses to blue light which have been detected recently (by the use of such sensitive methods of growth measurement as linear displacement transducers, see Ch. 6). However, various lines of evidence suggest that the response popularly termed 'blue light growth inhibition', is *not* the basis of phototropic curvature:

(1) In etiolated seedlings of cucumber, a strong (inhibitory) growth

response is apparent within minutes of exposure to blue light, but phototropic curvature takes several hours to develop (Cosgrove 1985).

(2) Conversely, in de-etiolated mustard seedlings, there is only a very weak growth response to blue light, but a strong phototropic response (Rich *et al.* 1985).

(3) If the light on one side of a bilaterally irradiated coleoptile is switched off, it is the growth of the remaining irradiated side which ceases (yet its light environment is essentially unchanged); this suggests that it is the actual light gradient across the organ that is important, rather than the absolute levels of light on each side (Macleod *et al.* 1985).

The last finding also serves as a reminder that tropic curvature involves co-ordination of growth responses on both sides of the organ. Moreover, curvature is rarely due entirely to phenomena of growth inhibition; in most cases, there is also significant growth stimulation on the shaded side of a unilaterally irradiated organ (Hart *et al.* 1982).

There are indications that the actual patterns of change in growth can differ in relation to the precise protocol for phototropic stimulation,

Key

+++, ++, + extent of stimulation

− growth inhibition

(a) Transfer from darkness to unilateral light

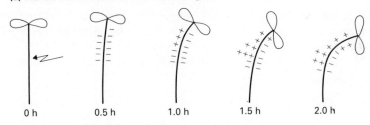

| 0 h | 0.5 h | 1.0 h | 1.5 h | 2.0 h |

(b) Transfer from overhead light to unilateral light

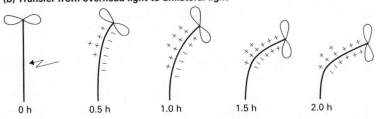

| 0 h | 0.5 h | 1.0 h | 1.5 h | 2.0 h |

Figure 7.12 Changes in growth rates of cress hypocotyls in response to unilateral irradiation. Different patterns of change are obtained with different protocols of phototropic stimulation.

even within the one species and even though the final curvatures appear similar (Fig. 7.12). In cress seedlings stimulated by transfer from darkness to unilateral light, curvature results from a temporary inhibition of growth on both sides of the hypocotyl, followed by recovery of growth first on the shaded side. Transfer from omnilateral to unilateral light, on the other hand, results in rapid simultaneous growth inhibition on the lit side and growth stimulation on the shaded side (stimulation, incidentally, to a greater extent than is induced by transfer to total darkness).

The involvement of both growth inhibition and growth stimulation in phototropic curvature has been pointed out several times. In 1937, Went and Thimann suggested that curvature in the coleoptile resulted from growth inhibition on the lit side near the tip of the organ, and growth stimulation on the shaded side towards the base. In 1969, Curry affirmed that first positive types of curvature were due to growth inhibition on the lit side, while second positive curvature derived from enhanced growth on the shaded side. However, most complex of all, in 1880 Darwin concluded from his painstaking observations that 'heliotropic movement manifestly results from a modification of ordinary circumnutation' in which 'the movement consisted of a succession of ellipses directed towards the light, each of which approached closer to the source than the previous one'. Apparently he envisaged a sequence of constantly changing growth rates on each side of the coleoptile, rather than either stimulation or inhibition alone on separate sides. Therefore, past observations and modern measurements seem to agree that the growth responses involved in tropic curvature are much more complex than the perhaps popular notion that simple growth enhancement is responsible for all types of curvature.

7.4 CONCLUDING REMARKS

There are several distinct mechanisms by which different types of plant can use light to attain a particular orientation: change in position of the apical growing point in moss protonemal filaments; asymmetric growth reactions in the fungal sporangiophore; differential cellular growth in higher plants. Within higher plants, there are different photoactivated mechanisms for achieving differential growth: red-light growth responses and blue-light phototropic responses. And even within blue-light induced photocurvature, differential growth across the organ can probably arise from different patterns of growth in different situations.

In the face of this great diversity of 'phototropic response', it is perhaps unfortunate that there has been such concentration on one

158

organ (the coleoptile), one light response (first positive curvature) and one transduction mechanism (an auxin gradient).

7.5 SUMMARY

(1) Tropic growth responses enable a plant to regulate and adjust its orientation. The major tropic stimuli are light and gravity, with chemicals and touch effective in certain situations.

(2) Phototropism in the oat coleoptile shows a typical blue light response action spectrum; there is a complex fluence-response relationship, with different types of positive curvature in response to different combinations of exposure period and fluence rate. The form of response known as second positive curvature is probably the form which occurs under natural conditions; its extent is a function of duration of exposure rather than fluence or fluence rate; and it can occur in the absence of the coleoptile tip.

(3) Dicot seedlings are much less phototropically sensitive than coleoptiles. Their curvature is mainly of the second positive type.

(4) Red light pretreatment changes the phototropic sensitivities of organs; red light itself seems capable of indirectly affecting the orientation of dicot seedlings through growth regulating effects transmitted from leaves or cotyledons: red light also directly affects the directional growth of moss protenemal filaments.

(5) The Cholodny–Went hypothesis is generally accepted as the explanation of how a directional light stimulus produces directional growth responses in higher plants. According to this hypothesis, tropic stimulation establishes an auxin gradient which leads to differential rates of growth across the tissue. Major difficulties with this hypothesis relate to the nature of the changes in growth rate responsible for curvature; these changes occur very rapidly, and involve patterns of growth inhibition and stimulation on each side of the curving organ. Furthermore, it seems as though different patterns of growth response are induced by different protocols of phototropic stimulation.

FURTHER READING

Curry, G. M. 1969. Phototropism. In *Physiology of plant growth and development*, M. B. Wilkins (ed.), 245–73. Maidenhead: McGraw-Hill.
Darwin, C. 1880. *The power of movement in plants*. London: John Murray.

Dennison, D. S. 1984. Phototropism. In *Advanced plant physiology*, M. B. Wilkins (ed.), 149–62. London: Pitman.
Went, F. W. & K. V. Thimann 1937. *Phytohormones*. New York: MacMillan.

CHAPTER EIGHT

Orientation in time: photoperiodism

8.1 INTRODUCTORY COMMENTS

Obvious features of plant behaviour are its periodicity and synchrony with environmental conditions – different activities are associated with particular times of the day and with certain seasons. What is not so obvious is that much of this periodicity is not imposed directly by the environmental conditions at that time. Plants, like all other eukaryotes, have the ability to anticipate many changes in the environment, inasmuch as certain signals induce sequences of development or patterns of behaviour which prepare the organism for future conditions. This applies both to daily and to seasonal cycles of change: plants seem to have both a clock and a calendar. Light is strongly involved in each of these aspects of time measurement.

Many of the fluctuations in a plant's activities which occur during the daily cycle are under endogenous control, and are said to exhibit a **circadian rhythm** (Table 8.1). Although light is not involved in the direct induction or operation of many of these activities, it interacts to a very significant extent with the mechanism responsible for their timing.

The major means by which plant development is synchronized with the seasonal cycle is through **photoperiodism**, where responses are induced by changing daylengths. At latitude 57°N (Aberdeen, Scotland; Gothenburg, Sweden; Anchorage, Alaska = 61°N), the range of daylengths over the year is 7–18 hours, and the rate of change in daylength (spring and autumn) is around 38 minutes per week. Since many plants can discriminate between differences in daylength of 38 minutes, this means that their photoperiodic responses can be timed to within the same week, year after year. The ranges of daylength and the rates of change are, of course, greater at higher latitudes (Fig. 3.5), and it is often thought that photoperiodism is therefore restricted to plants of temperate regions. However, many photoperiodic responses are known in tropical plants. Indeed, such responses indicate a very

Table 8.1 Examples of processes exhibiting a circadian rhythm.

Organism	Process	Spectral region involved in timing
higher plants	leaf position flower movements	blue, red, far-red
	stomatal movements cell division growth rate respiration photosynthesis	blue, red
	dark CO_2 fixation root pressure many enzyme activities gene expression	red
Gonyaulax	bioluminescence photosynthesis cell division	blue, red
Neurospora	conidiation	
Pilobolus	spore discharge	
Euglena	phototaxis	
Drosophila	pupa emergence	blue
man	body temperature urine production blood sugar level	

high degree of daylength discrimination: at latitude 30°, the daylength range is 10 to 14 hours and the rate of change is only 12 minutes per week.

Although photoperiodism is frequently associated with the *initiation of flowering*, a great many other areas of plant development are also subject to photoperiodic control (Table 8.2). These include other aspects of floral reproduction, such as *fertility* (e.g. pollen viability in cocklebur and actual flower development in chrysanthemum) and *sex determination* (e.g. in maize, more female flowers develop if short days are experienced early in the reproductive phase). (Plant sexuality can also be influenced by light in a non-photoperiodic way: in many cases, the production of female flowers is enhanced by high photon fluence rates, although in the shade plant *Mercurialis* (dog's mercury), more male plants are found in habitats of high fluence rates.) Another major area of photoperiodic response is the induction of bud dormancy, particularly in woody species. Generally, shortening daylengths induce bud formation. (Bud break in the spring is more commonly controlled by temperature, but in a few species it occurs in response to long

162

Table 8.2 Examples of photoperiodic control in plant development. (summarized from Vince-Prue 1975).

Aspect of development	Process	Stimulus	Example
floral reproduction	flower initiation	SD, LD, S–LD, L–SD	see text
	flower development	continued SD	*Chrysanthemum*
	fertility (viable pollen)	continued SD	*Xanthium*
	sex expression (female flowers)	SD	maize
vegetative perennation	induction of bud dormancy	SD	woody perennials
	runner development	LD	strawberry
	development of storage organs	SD	tubers, bulbs, corms
	development of cold hardiness	SD	*Robinia*
	leaf abscission	SD	apple
growth habit	stem elongation	LD = taller	tomato, bean
	leaf growth	LD = larger, thinner	*Impatiens*
	branching	SD = more branching	many grasses
	root initiation	LD = more initiation	*Bryophyllum*

Note: SD, short days; LD, long days; S–LD, short days followed by long days; L–SD, long days followed by short days.

days.) Other methods of vegetative reproduction and perennation are also photoperiodically controlled – for example, the formation of most forms of *storage organ* is induced by short days. *Plant form* is strongly influenced by photoperiod, through effects on stem growth, branching pattern, apical dominance and leaf development.

The adaptive significance of many photoperiodic responses has still really to be considered. However, control over the timing of its various developmental phases would seem to be advantageous to a plant in two general ways, the relative importance of which may differ in different situations:

(1) *Synchrony with the physical environment*. Particular stages of development can be phased with appropriate climatic conditions – for example, vegetative development with favourable growing conditions. (This aspect may be particularly relevant to plants of temperate latitudes where there are wide ranges of climate over the year.)

(2) *Synchrony with other individuals and organisms*. The timing of certain activities in relation to the behaviour of other individuals and organisms could be of advantage in several respects; if

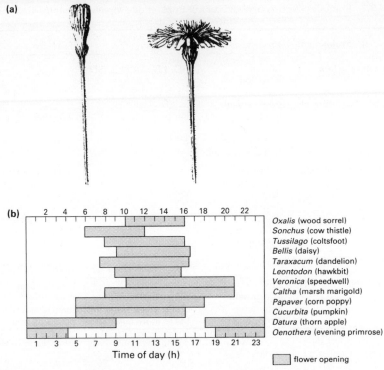

Figure 8.1 Photonastic movements in flowers (a) Flower scape of *Leontodon hastilis* in the night (left) and day positions. (b) Times of flower opening and closing in various species.

individuals of the same species enter their reproductive phases at the same time, the possibility of outbreeding will be increased; if individuals of different species (e.g. a plant and its pollinators) also synchronize their activities, reproduction could be enhanced; and temporal control of development could allow the same ecological niche to be occupied by different organisms at different times. (It is these aspects which may be of greater significance in regions where seasonal differences in climate are not so large.)

8.2 CIRCADIAN RHYTHMS

8.2.1 Introduction and terminology[1]

Many activities in an organism show a regular periodicity. Different types of rhythm can be distinguished by the lengths of a single cycle.

[1] See Mansfield & Snaith (1984), Palmer (1976), Sweeney (1977, 1979).

These include daily, or *circadian*, cycles of activity, and others with cycle lengths of hours (*ultradian*, e.g. nutational growth movements) or a month (*lunar*, e.g. some responses in littoral organisms) or a year (*circannual*, e.g. some types of germination). Circadian rhythms are well illustrated by the regular, daily movements of certain leaves and flowers. Although the phenomenon of 'plant sleep' in *Tamarindus indica* (the tamarind) was described in the fourth century BC by Androsthenes, one of the generals of Alexander the Great, such movements were first the subject of serious investigation in 1729 by De Mairan, who showed that continuation of the daily cycle of leaf movement in beans was not dependent upon continuing exposure to an environmental cycle of light and darkness. (In a more practical vein, Linnaeus constructed a 'flower clock' by noting the times of flower movements in various species, like that in Fig. 8.1; his idea was that this would be useful to people walking in the countryside without a timepiece.) Many other responses in both plants and animals also show a daily periodicity (Table 8.1). No circadian rhythms have been observed in prokaryotes.

Use of the term 'circa-dian' ('about a day') stems from the fact that when all the environmental conditions are held as constant as possible – when the biological rhythm is therefore *free-running* – the length of one cycle of activity, or *period* (Fig. 8.2), is not exactly 24 hours. *Entrainment* describes the setting of the rhythm; under natural conditions it involves bringing the period into synchrony with

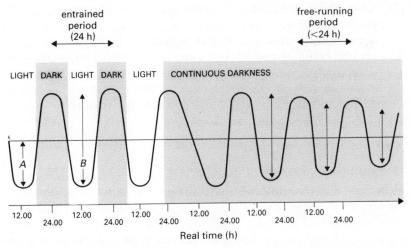

Figure 8.2 The major characteristics of a circadian rhythm. During daily cycles of light and dark, the *period* is *entrained* to 24 hours; under constant conditions (darkness), the *free-running* form of the rhythm becomes established and the period is not 24 hours; in this example, the *phase* has also been delayed by the extension of the third light period before continuous darkness; in constant darkness, the *amplitude* (A) and the *range* (B) become *damped*.

165

the 24-hour daily cycle. Light is involved in the entrainment of many rhythms, and both dawn ('light on') and dusk ('light off') can act in different rhythms as entrainment signals, or *zeitgebers*. The *phase* describes the state of the cycle at any one point in time; a *phase shift* occurs if the cycle changes in relation to the time axis (Fig. 8.2). In this connection, the terms 'circadian time' or 'subjective time' are also often used to describe the endogenous stage that the cycle has reached, as distinct from external or objective time; phase shift thus involves an initial dislocation between circadian time and external time. The *amplitude* of the cycle describes the extent to which the response deviates from its mean value, and the *range* indicates the difference between the maximum and minimum values (Fig. 8.2). These last two features, the amplitude and the range, can be markedly affected by many environmental factors, but it is important to realize that these only represent effects upon the extent to which the rhythm is expressed, not upon its timing or periodicity. Reduction in the amplitude, *damping*, usually occurs when the rhythm is free-running under constant conditions (Fig. 8.2).

Before a regular pattern of behaviour is accepted as being controlled by some endogenous mechanism, certain criteria must be satisfied (see Mansfield & Snaith 1984):

(1)   the regularity must continue under constant environmental conditions (which usually involve continuous darkness);
(2)   when the rhythm is free-running under constant conditions, the periodicity should not remain fixed at exactly 24 hours at all temperatures (this avoids the possibility that the response is simply being triggered by some unsuspected environmental regularity);
(3)   it should be possible to phase-shift the rhythm by an abrupt change in some environmental factor (often, light).

8.2.2   Effects of temperature[2]

A notable feature of circadian rhythms is their response to temperature. The period of a free-running rhythm remains fairly constant over a wide range of temperatures. This is the kind of behaviour that is, of course, required of a timing mechanism: that it be uninfluenced by fluctuations in environmental temperatures. (Note, however, that the amplitude, i.e. the extent of expression of the rhythm, is usually very much affected by temperature.) The relative constancy of period at

[2] See Mansfield & Snaith (1984), Palmer (1976), Salisbury & Ross (1978).

different temperatures is often referred to as 'temperature independence' of the rhythm, which suggests the rather unlikely situation of a biological process being unaffected by temperature. A more accurate description may be that circadian rhythms show *temperature compensation*. This is meant to suggest that rhythm timing results from the actions of more than one process, each of which has a different temperature coefficient, and that temperature compensation is achieved through some sort of balanced response.

Circadian rhythms in plants generally do not show as complete temperature compensation as in animals. For example, the free-running period in bean leaf movements is somewhat influenced by temperature:

Temperature	Period
15°C	29 hours
25°C	24 hours
35°C	19 hours

Note that in this case, an increase in temperature results in a decrease in cycle length. There are also cases where increase in temperature gives an increase in cycle length, as with the rhythmic bioluminescence in the marine dinoflagellate, *Gonyaulax polyedra* (Salisbury & Ross 1978).

8.2.3 Effects of light[3]

Light is the environmental factor which has the most significant effects on the timing of circadian rhythms, certainly with regard to the majority of those in plants. It can have four kinds of effect on a rhythm: treatment with light can change the phase, stop the rhythm, influence the period and entrain the rhythm.

Brief exposure to an appropriate type of light at particular times during a circadian cycle can induce phase shift, especially if the light treatment is given during the subjective dark stage of the cycle. It is possible to produce a **phase-response curve** (Fig. 8.3) by plotting the time in a circadian cycle at which the light treatment is given (abscissa) against the extent of phase shift that is brought about (ordinate). The time of treatment not only affects the amount of phase shift, but also its direction: if the light treatment is given early in the dark period ('false dusk'), the rhythm is delayed (see Fig. 8.3); if it is given late in the dark period ('false dawn'), the rhythm is advanced. Obviously, there is also a stage of treatment when rhythm delay suddenly

[3] See Sweeney (1977, 1979).

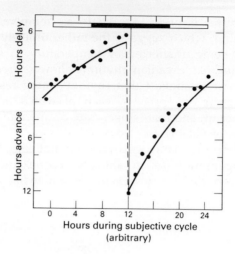

Figure 8.3 The response curve for phase-shift of a circadian rhythm by light treatment. The extent of delay or advance in the phase of the rhythm is plotted against the point in the rhythm at which the light treatment was given (the bar at the top indicates the subjective status of the rhythm, i.e. the dark part indicates subjective night). This example is for change in the rhythm of petal opening in *Kalanchoe* in response to a 24-hr treatment with orange light.

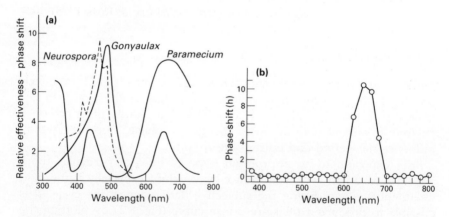

Figure 8.4 Action spectra for phase-shift of circadian rhythms in different organisms. (a) Approximate spectra for phase shift in *Neurospora*, *Gonyaulax* and *Paramecium*. (b) Spectrum for phase shift in the rhythm of CO_2 output in *Bryophyllum*.

becomes rhythm advance. This same form of phase-response curve is seen in the effects of light on the rhythms in many different types of organism.

Action spectra for light-induced phase shifts, however, are different for responses in different types of organism (Fig. 8.4), and can even differ between responses in the same organism. This indicates that

168

different photosystems interact with the circadian timing mechanism. In many animals and most lower organisms, the blue region of the spectrum exerts the strongest effects. In higher plants, blue and red light are active, jointly or individually depending upon the particular rhythmic response being considered: for example, both blue and red light induce phase shift in the rhythm of leaf movements in legumes; but only red light is effective on the rhythm of dark CO_2 fixation in *Bryophyllum* (Fig. 8.4); and in the rhythmic leaf movements of *Coleus*, treatment with red light advances the phase, while blue light delays it (Mansfield & Snaith 1984). The negation of red-light induced phase shifts by far-red irradiation implicates phytochrome in many rhythmic phenomena in higher plants.

Other effects of light on circadian rhythms include the action of continuous bright light in halting many rhythms. (This is in contrast to the development of the free-running form of the rhythm in continuous darkness.) Further, the period of many rhythms shows some dependency on light intensity. This latter effect has been oberved particularly in the rhythms of animals, and is sometimes referred to as Aschoff's rule: high light intensities lengthen the rhythm period in diurnal animals and shorten the period in nocturnal animals (Sweeney 1977). The action spectra and mechanisms of these effects are not known.

Light also has very significant effects on **rhythm entrainment**. In fact in plants, light is the major rhythm entrainment or clock-setting agency. (It is worth reminding ourselves at this point that light is the most regular and reliable oscillating factor in the environment.) Under natural conditions, exposure to the regular cycles of light and darkness entrains the free-running circadian rhythm to a 24-hour period. Exposures to differently timed artificial regimes of light and dark can give entrainment to periods longer or shorter than the natural 24 hours. Entrainment is thought to occur through processes of phase shift. That is, if the external cycle of light and darkness does not match the circadian cycle, the rhythm will receive 'unaccustomed' light either early in the circadian dark period (causing phase delay) or late in the circadian dark period (causing phase advance). Successive adjustments of phase will quickly result in a match between the circadian cycle and the external cycle.

8.2.4 *The nature of the endogenous clock*[4]

The different rhythms in an organism are thought to be controlled by a single timing mechanism, but the actual nature of this cellular clock represents one of the major unsolved problems in biology.

[4] See Mansfield & Snaith (1984), Njus *et al.* (1974), Sweeney (1977).

There is a certain amount of circumstantial evidence that the timing mechanism (somehow) involves cell membranes. Many agencies that affect the properties of membranes also produce a phase shift in some circadian rhythms. (Note that it is an effect on the timing of the rhythm that is important, not merely an effect on the extent of a response.) In some cases, these agencies are extremely non-specific; for example, treatment with low temperature alters the phase of rhythmic chlorophyll fluorescence in *Gonyaulax* (by suggested effects on membrane fluidity); again, rhythms in higher plants are altered as a consequence

Box 8.1   The practical significance of circadian timing

The circadian timing mechanism is of great practical significance (Palmer 1976). It is, of course, significant to the organism itself, in forming the basis for its *temporal orientation* on both a daily and, as we shall see, a seasonal scale. (Besides this obvious role, circadian timing is also important for the *spatial orientation* of some organisms. Direction-finding in birds and insects is based upon perception of the pattern of polarization in reflected skylight; since this pattern changes regularly throughout the day and with the season, its interpretation requires the organism to have an extremely precise awareness of time.)

However, circadian rhythms in plants are of significance also to man, in at least two respects. Their operation means that in many (most?) situations, response to a treatment will be highly dependent upon the time at which the treatment is given. Therefore, in the first place, the actual timing of certain agricultural practices or agrochemical treatments could be commercially significant – for example, in getting a particular level of response from a lower dose at a lower cost. (It is well established that the effects of many drugs on animals, and of alcohol on man, vary according to the time of day at which they are administered.) In the second place, the existence of circadian rhythms should be taken into account in experimentation with plants. It can be important to time an experimental treatment with due regard to the rhythms already established by previous cycles of light and darkness (or even by wilting). An example of this is seen in the behaviour of stomata (Fig. 8.5). Guard cells operate under a strong circadian rhythm. The stage of the cycle at which a treatment is applied has a great effect on the extent of response; in the case illustrated, the response of the guard cells to light is highly dependent upon the length of the preceding dark period. There will obviously be 'knock-on' effects from this to other areas, such as rate of photosynthesis.

It has been recommended many times that experimentation be carried out on light–dark cycles identical to those under which the plant material has been prepared; this may involve deliberately setting the phase to coincide with available experimental time.

of wilting (mentioned here primarily for its practical import). Treatment with chemicals such as ethanol, ether and caffeine also induce phase shifts, although again the specificity of action could be questioned. However, the action of the ionophore, valinomycin, in producing phase shifts in the rhythms of *Gonyaulax* and of bean leaf movements, suggests more specifically that the transport of K^+ across a cell membrane is somehow involved in biological time measurement (see Mansfield & Snaith 1984). (An ionophore markedly enhances the permeability of a membrane to a particular ion).

It has been proposed that the mechanism of time measurement may involve a membrane-located ion pump that is de-activated by a feedback loop. The idea is that the pump would actively transport a particular ion into some cellular compartment until it was switched off by the resulting high concentration of ion itself. The ion would then begin to passively leak out of the compartment – and it is this process which would act as the timing mechanism. Therefore, any factor which interfered with ion outflow, either by damaging membrane permeability or by activating the pump at the 'wrong' time, would also interfere with timing. Present views, of course, attribute at least some aspects of phytochrome action to modification of membrane function (Ch. 5). Thus, phytochrome may interact with the biological timing mechanism through activation of the ion pump: a light exposure early in the circadian dark period would delay timing, through activation of the pump and consequent delay in the onset of passive leakage; a light signal late in the dark period would advance timing by stopping passive leakage at an earlier point.

Such ideas carry the clear indication that *phytochrome itself is not the circadian clock*. The rôle of phytochrome, and of other photosystems which interact with biological rhythms, is to link the regular periodicity

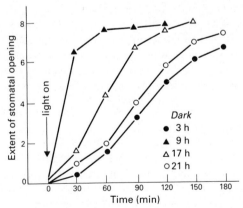

Figure 8.5 Rate of stomatal opening in response to light after different periods in darkness.

of the (external) light environment to the (internal) cellular timing mechanism.

Some of the more practical aspects of circadian rhythms are discussed in Box 8.1.

8.3 PHOTOPERIODISM I: EARLY STUDIES

8.3.1 Historical basis and plant types[5]

The discovery of photoperiodism is generally recognized to mark the start of research on photomorphogenesis (in which case, the study of the effects of light on plant development was initiated by placing plants in absolute darkness!). The actual discovery of photoperiodism is credited to Garner and Allard in 1920, although a few previous investigators had suggested that daylength was important to plant development. Indeed, in 1914, the Frenchman, Tournois, wrote that flowering in a short-day plant was influenced by the length of the dark period.

Garner and Allard were plant breeders – hence their interest in obtaining flowers. They had difficulty in getting two particular species to flower: a variety of tobacco, Maryland mammoth, only flowered if grown in winter (indoors); and soybean (*Glycine max*) only flowered in late summer, no matter at what time it was sown during the growing season. After fruitlessly investigating the effects of other environmental factors (such as light intensity, temperature, nutrition), Garner and Allard realized that daylength was also an important seasonal variable. They tested the effects of short daylengths by growing the two species in a hut where daylength could be artificially shortened in summer (the site of this hut is somewhere inside what is now the Pentagon complex). The successful induction of flowering in both species indicated that daylength was indeed an important factor in the regulation of plant development. (It was five years later, in 1925, before photoperiodism was reported in birds and insects.)

Investigation of many other species resulted in the initial recognition of three classes of plant, distinguished by the effects of daylength on flowering (Table 8.3):

(1) *Short-day plants* (SDP): generally species from low latitudes (e.g. coffee, cotton, rice), or species which flower in late summer, such as chrysanthemum (see Cockshull 1984).

(2) *Long-day plants* (LDP): generally species from high latitudes, such as temperate grasses (see Deitzer 1984).

[5] See Salisbury & Ross (1978), Thomas & Vince-Prue (1984), Vince-Prue (1983a).

Table 8.3 Short-day, long-day and day-neutral flowering plants.

Daylength requirement	Degree of control	Species
short-day plants	obligate (absolute requirement)	chrysanthemum
		coffee
		poinsettia
		strawberry
		tobacco (var. Maryland mammoth)
		duckweed (*Lemna*)
		cocklebur (*Xanthium*)
	facultative (quantitative requirement)	hemp (*Cannabis*)
		cotton
		rice
		sugar cane
long-day plants	obligate	carnation (*Dianthus*)
		henbane (*Hyoscyamus*)
		oat (*Avena*)
		ryegrass (*Lolium*)
		clover
	facultative	pea
		barley
		lettuce
		wheat (spring wheat)
		turnip (*Brassica rapa*)
day-neutral plants		cucumber (*Cucumis*)
		Fuchsia
		tomato (*Lycopersicum*)
		potato
		bean (*Vicia faba*)
		dwarf bean (*Phaseolus vulgaris*)
		rose

(3) *Daylength neutral* (DN): usually species with a wide latitudinal distribution, such as potato, tomato (see Halevy 1984).

This original classification has since become considerably more complex with the introduction of various subclasses (Table 8.3). For example, plants differ in respect of the strictness of dependence on daylength, showing:

(1) *Obligate* (or qualitative) photoperiodism: where there is an absolute requirement for a particular daylength (SD or LD).
(2) *Facultative* (or quantitative) photoperiodism: where a particular daylength advances or enhances flowering, but the plants will eventually flower anyway.

Some species require a regime of two types of daylength:

(1) *Short days to long days* (SLDP): some spring flowers (e.g. white clover).

(2) *Long days to short days* (LSDP): some autumn flowers (e.g. *Bryophyllum*).

However, such classifications are to a large extent empirical, since within a single species there can be ecotypes with different photoperiodic requirements. (This also contributes to the wide latitudinal distribution of a species, with LDP varieties at the higher latitudes and SDP varieties at the lower latitudes.) There is also great variation in photoperiodic sensitivity between species. This is seen both in regard to the degree of daylength discrimination (sensitive species like rice and sugarcane can discriminate between differences in daylength of 15 minutes) and in terms of the numbers of appropriate light–dark cycles that are required to induce flowering. (It is worth bearing in mind that most of our information on photoperiodism comes from experimentation with 'convenient' species which require only one (e.g. *Xanthium, Pharbitis, Chenopodium, Lemna*) or a few (e.g. *Perilla, Glycine*) inductive cycles). Another point to note is that a species can be day-neutral in respect of flowering, but photoperiodically responsive in other respects (e.g. 'day-neutral' tomato is bushy under SD and tall under LD).

It is important to realise that the distinction between SDP and LDP lies, not in any requirement for a particular daylength, but in their differing responses to changing daylength. This is illustrated in Figure 8.6, where the idealized flowering responses for these two types of plant are graphed against daylength:

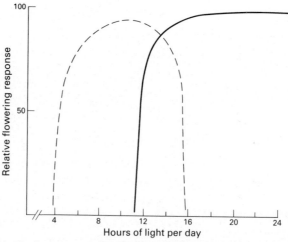

Figure 8.6 Idealized representation of the flowering responses of short-day plants (– – – –) and of long-day plants (————) to different daylengths.

(1) In LDP, as the light period is reduced to less than 13 hr, flowering decreases – that is, LDP require a daylength of more than some critical minimum.

(2) In SDP, as the light period is increased to more than 11 hr, flowering decreases – SDP require a daylength of less than some criticial maximum.

Two other important features are also apparent in Figure 8.6:

(3) SDP do need some period of light (in most SDP, flowering decreases at very short daylengths).

(4) There is overlap between the *critical points* of the two types (both types diagrammed here would flower at 13 hr).

The last point reinforces the statement that the SDP vs. LDP distinction does not represent a requirement for a specific daylength, in fact, the critical points differ not only between species, but can even change within a species as a consequence of changes in age, temperature or nutrition (see Box 8.2).

The perception of the photoperiodic stimulus occurs in the leaves and results in some physiological transformation of the leaf material into what is termed the 'induced state'. Production of this induced state also shows some variation between species: in *Xanthium*, exposure of about 2 cm^2 of leaf material to an SD stimulus results in all the leaves of that plant becoming induced; in *Perilla*, only the actual tissue exposed to the photoperiodic stimulus becomes induced. In both these species, the induction is permanent; in soybean, reversion to the non-induced state is possible. Virtually nothing is known about the nature of the induced state, or, indeed, about any part of the photoperiodic transduction chain. It is from the induced state of the leaves that some 'flowering stimulus' is transmitted to the stem apex to 'evoke' the changes that result in the production of floral primordia. Although a great deal of effort has been expended in this area, the nature of the flowering stimulus is completely unknown; it may be a mixture of substances rather than a single chemical – or the 'stimulus' may even be the removal of a flowering inhibitor.

8.3.2 Light–dark sensitivity[6]

An important initial question in photoperiodic research concerned the actual feature of the daily period that was responsible for stimulation – whether it was the light period, the dark period or some ratio of

[6] See Salisbury & Ross (1978), Vince-Prue (1983a).

Box 8.2   Other controls of flowering

Response to photoperiod can be so dramatic that it is often overlooked that there are other controls of flowering (Thomas & Vince-Prue 1984). Much flowering occurs without any photoperiodic involvement, and even many photoperiodic species eventually flower without daylength control. Indeed, species are often photoperiodic only under certain conditions – for example, the classic Maryland mammoth is photoperiodic under high summer temperatures, but is day neutral at temperatures lower than 12°C.

Other parameters involved in the regulation of flowering include endogenous factors, such as *age*, and external factors, particularly *temperature* and *nutrition* (e.g. a low nitrogen regime is used commercially to induce flowering in tomato and apple). Species differ in the extents to which particular factors are important. In some cases, age is the only controlling influence; flowering in these plants is said to be 'insensitive' to the environment.

The age at which a plant flowers, or becomes 'sensitive' to environmental induction of flowering, is related to the concept of *juvenility* – the juvenile phase being the period of growth during which no flowering is possible; it is often also characterized by morphological differences (e.g. the leaf shape, phyllotaxis and growth rate). The duration of the juvenile phase varies greatly: a few species can flower at the cotyledonary stage (e.g. *Pharbitis, Arabidopsis*); others take much longer (e.g. bamboos need 40 years). Woody plants in general have a long vegetative phase before becoming sensitive to environmental control (e.g. birch 5–10 years, and beech 30–40 years). (Then, conversely, the older a plant gets, the less strict become the conditions required for flowering, until some species eventually become 'insensitive'.) Juvenility is thought to be a device to ensure that a plant is large enough to support the demands of seed production. The actual changes that occur when a plant passes from the juvenile to the 'ripe to flower' stage are not clear, although the size of the meristem seems to be important.

Temperature is another environmental factor which shows regular daily and seasonal fluctuations, though not so precisely as light. These fluctuations in temperature also have major effects on both plant growth and flowering. In the first place, almost all plants grow better when the night temperature is lower than the day temperature (thermoperiodism). Even more notable are the effects of low temperatures on the flowering of some species. The phenomenon is called *vernalization*: treatment of seeds or seedlings for 10–100 days at 0–10°C promotes flowering (see Napp-Zinn 1984). A requirement for vernalization confers upon a plant the ability to obtain more (confirmatory) information about the season. That is, daylength alone can be a rather ambiguous signal in relation to spring or autumn; a preliminary requirement for vernalization means that winter must actually pass before the daylength signal is meaningful. The phenomenon was noted by Gassner, in 1918, in winter varieties of

176

cereals (in these, ear formation is delayed until after more than 20 leaves have been formed if seeds are sown in the spring and thus do not receive a low temperature treatment); it is also responsible for the behaviour of biennials which flower in their second year of growth (cabbages, carrots, beets), and is a control of flowering in some herbaceous perennials (primrose, wallflower). Many LDP remain in a vegetative rosette stage even under photoperiodically inductive conditions unless they have first been vernalized (temperate grasses). As little is known about the transduction chain and the metabolic changes induced by vernalization as about those induced by photoperiodic stimulation. The effect is perceived in the apex rather than the leaves, and seems to involve recently meristematic cells. The vernalized state is then a property of all subsequent daughter cells, but is dissipated upon sexual reproduction.

light : dark. By varying the lengths of the light and dark periods independently within a 24-hour cycle, Hamner and Bonner demonstrated in 1938 that it is the absolute length of the dark period that is critical: SDP require a dark period of more than some critical minimum length. This concept was reinforced by the results of what are called **night-break experiments**. That is, interruption of the long dark period by a brief light treatment inhibits flowering in SDP; it also induces flowering in (a few) LDP. The general situation is summarized in the diagrams of Figure 8.7.

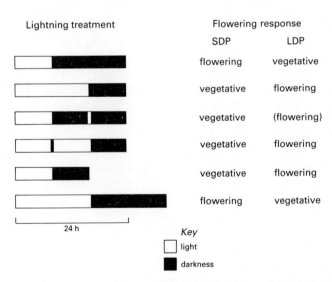

Figure 8.7 Diagrammatic representation of the effects of daylength and of light interruption of the dark period on the flowering responses of short-day plants and long-day plants.

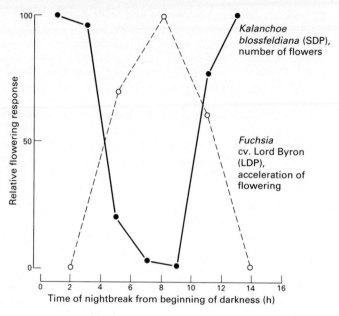

Figure 8.8 Effects of light-interruption at different times during the dark period, on flowering in a short-day plant and in a long-day plant.

Also in the early years of research into photoperiodism, determinations of action spectra ·and the results of R–Fr reversibility tests in night-break experiments indicated that phytochrome is the photoreceptor for these light effects in both SDP and LDP. In fact, light-interruption of the photoperiodic dark period is a classical example of the LER (Ch. 4), with relatively fast escape times of 2 minutes (*Pharbitis*) to 90 minutes (*Chrysanthemum*) for far-red reversibility, even though flowering responses are not usually expressed until several weeks later.

A major feature of the night-break type of experiment is that for maximum effect, the light interruption must be given at a certain stage in the dark period, usually around eight or nine hours from the end of the light period (Fig. 8.8). When this is considered along with the fact that in SDP, a certain minimum period of light is also required, an important conclusion emerges: in SDP, the induction of flowering is inhibited by light (P_{fr}) at some stages (night break), but actually requires light at other stages (light period). That is, we conclude this section with the realization that in photoperiodism, *the responses to light are different at different stages of the daily cycle.*

Some of the other factors besides light that also influence flowering were discussed in Box 8.2.

8.4   PHOTOPERIODISM II

8.4.1   *Interaction with endogenous rhythms*[7]

Photoperiodism is a phenomenon which obviously involves time measurement by some kind of clock. Since it is the absolute length of the dark period that is important rather than the ratio of light : dark periods, this clock must be capable of measuring absolute time rather than relative time. As the major photoreceptor which interacts in plant photoperiodism, phytochrome is somehow involved in this time measurement.

An early idea was that the physiological dark reactions of phytochrome provided the actual timing mechanism, through decay of P_{fr} down to a level that no longer interfered with certain 'dark-requiring' processes. However, it soon became apparent that this so-called *hourglass hypothesis* was not consistent with many physiological observations:

(1)   In the cases of *Xanthium* and *Chenopodium* the lengths of the

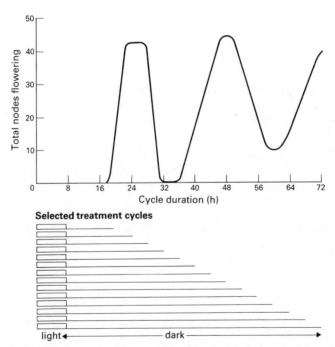

Figure 8.9   A resonance type of experiment on photoperiodic induction of flowering in soybean (SDP). Plants were exposed to cycles consisting of 8 hrs light and dark periods of different lengths; the extent of flowering is plotted against the total length of a cycle.

[7] See Thomas & Vince-Prue (1984), Vince-Prue (1983a).

critical dark period, i.e. time measurement, are not greatly affected by temperature, but the dark transformations of P_{fr} are highly dependent on temperature.

(2) The length of the critical dark period, and the time of maximum sensitivity to a night break, are not affected even if the level of P_{fr} is lowered by end-of-day far-red irradiation (the extent of flowering may be affected, but time measurement is not).

(3) Generally, the behaviour of P_{fr} does not coincide with the lengths of critical dark periods; in certain plants like *Lolium* and dark-grown *Pharbitis*, P_{fr} decay seems to be complete within a few hours; alternatively, in light-grown *Pharbitis*, the level of P_{fr} remains stable in darkness for many hours.

Therefore, *phytochrome itself is not the photoperiodic clock*. Present views are that photoperiodic time measurement is based on some sort of rhythmic process and that phytochrome interacts with this rhythm.

We have already seen that the effects of light on the photoperiodic system depend upon the stage at which the light is given (e.g. Fig. 8.8). In 1936, on the basis of his studies on circadian rhythms, Erwin Bünning suggested that photoperiodic sensitivity also operated on a circadian rhythm – the **Bünning hypothesis**. He suggested that the plant oscillates between two physiological states on which light has opposite effects: exposure to light during the *photophil* ('light-loving') phase has a positive, stimulatory effect; light exposure during the *skotophil* ('dark-loving') phase results in inhibitory effects. In 1958, Konitz demonstrated that there were indeed opposite responses to red light at different stages of a plant's cycle as determined by the nyctinastic position of the leaves: when the leaves of *Chenopodium amaranticolor* were oriented upwards (photophil), exposure to red light stimulated flowering but when the leaves were downwards (skotophil), red light inhibited flowering.

Four different types of experiment which indicate the existence of a rhythmic component in photoperiodic behaviour have been categorized by Vince-Prue (1983a):

(1) *Resonance experiments*. A short constant light period is coupled to dark periods of different lengths. When the extent of flowering is plotted against the total cycle lengths of light period plus dark period (Fig. 8.9), it can be seen that maximum flowering occurs when the total cycle lengths are some multiple of 24 hours. That is, flowering occurs when the light–dark cycle coincides with some (internal) rhythm. (This type of response has been demonstrated in both SDP and LDP, although there are some

species like *Xanthium* which do not show a clear rhythm under this treatment.)

(2) *Perturbation experiments*. A light break is given at various times during an extra-long dark period which extends over several normal daily cycles. Again a rhythmicity is apparent, this time in the degree of light-sensitivity (Fig. 8.10). (There are certain plants in which such a response is not straightforwardly apparent. In *Pharbitis*, a long preliminary photoperiod prevents the expression of rhythmic behaviour in the subsequent dark period, but appropriate conditions (Fig. 8.11) allow the demonstration of rhythmic fluctuation in light sensitivity.)

(3) *Skeleton photoperiods*. The 'photoperiod' consists of brief light exposures usually of 15 minutes, given at points equivalent to the beginning and end of a full photoperiod – for example, the sequence 15 mins L + 10.5 hrs D + 15 mins L : 13 hrs D is a skeleton photoperiod equivalent to 11 hrs L : 13 hrs D (Fig. 8.12). In the SDP *Lemna perpusilla*, such short-day skeleton photoperiods can substitute for complete light periods, again suggesting that the photoperiodic system can respond to coincidence between the external light environment and some internal oscillation.

(4) *Phase-shift experiments*. The rhythmic sensitivity to light can be advanced or delayed by exposure to light at appropriate

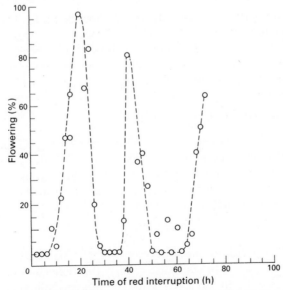

Figure 8.10 A perturbation type of experiment on flower induction in *Chenopodium rubrum*. Plants were given a 72-hr dark period, interrupted by 4 mins red light at different times.

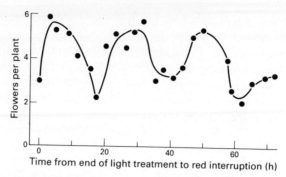

Figure 8.11 Demonstration of a rhythm in photoperiodic sensitivity in flowering in *Pharbitis*. Seedlings received a 10-min. photoperiod and a 72-hr dark period (and were also treated with benzyladenine). A rhythmicity is apparent in the sensitivity to a 10-min. night-break given at various times during the inductive dark period.

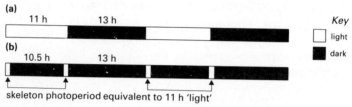

Figure 8.12 Diagrammatic representation of light regimes in (a) normal short-day photoperiods (b) short-day 'skeleton photoperiods' achieved by light pulses of 15 minutes.

circadian times in the plant cycle – for example, light early in the dark period delays the rhythm. That is, the rhythm of light-sensitivity in photoperiodism shows a typical phase-shift response curve analogous to those obtained for phase shifts in circadian rhythms (Fig. 8.13).

8.4.2 Photoperiodic time measurement[8]

It is now generally accepted that circadian timing is involved in photoperiodic behaviour in plants. There is a rhythmic sensitivity to light, which is expressed as phases when light (P_{fr}) is required for flowering, and phases when light (P_{fr}) is inhibitory to flowering.

There are two models which attempt to account for photoperiodic time measurement (Vince-Prue 1983a):

(1) *External coincidence model.* According to this model, a positive photoperiodic response arises when the (external) light cycle in the environment coincides with the appropriate stages of some (internal) plant rhythm; thus in this model, light is considered to

[8] See Vince-Prue (1983a), King (1984).

182

be involved in two distinct types of process; it sets the phase of the rhythmic sensitivity (to light itself), and it also has another direct effect on the induction of flowering during the light-sensitive (photophil) stage of the rhythm.

(2) *Internal coincidence model.* In this model (developed by C. S. Pittendrigh from studies mainly of animal systems) light is involved only in the timing mechanism; two distinct rhythms are considered to be entrained by light-off and light-on signals respectively; a positive photoperiodic response is thought to arise when appropriate points in these two (internal) rhythms coincide; thus in this model, light does not play a direct rôle in flower induction itself, but is dually involved in phase setting.

The external coincidence model is, of course, a development of the original Bünning hypothesis, and many recent accounts of plant photoperiodism tend towards this view. Photoperiodic time measurement in flowering is considered to begin at the 'light-off' point, and to continue through the dark period. That is, dusk initiates (or releases) the endogenous rhythm which determines the period during which light is inhibitory (the 'critical dark period' of the older literature). Actual photoperiodic induction is determined by the timing of dawn in relation to the stage of this rhythm: if 'light-on' occurs at the correct stage, the induced state is brought about. The light period, even in SDP, is thus necessary in two respects: it is required to entrain the rhythm of photoperiodic sensitivity; and it is required for the actual

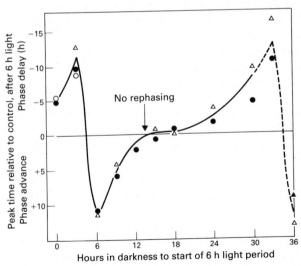

Figure 8.13 Response curve for phase-shift of photoperiodic timing in *Chenopodium*. Plants received 6-hr red light at different times in a free-running rhythm of flowering.

induction of the flowering response. Phytochrome seems to be involved in each of these aspects. It is involved in entraining circadian timing (a function fulfilled by different photoreceptors in other responses). But, in this model, it is also considered to be specifically involved in the induction of flowering.

Some investigators, however, consider 'external coincidence' to be invalidated by certain types of behaviour, and favour a form of internal coincidence model (King 1984). For example, *Pharbitis* flowers after only a single inductive light period – that is, it responds to one 'light-off' signal without any need for coincidence with a further dawn signal.

The actual basis of photoperiodic time measurement is a strongly debated area, with no generally accepted model.

8.4.3 Long-day plants[9]

Rhythmic responses are also apparent in the interactions of phytochrome with the photoperiodic behaviour of long-day plants – for example, there are rhythmic changes in the sensitivities of many LDP towards red and far-red radiation, or towards the R : Fr ratio, during the light period. However, most LDP do *not* seem to be either 'simple' mirror images of, or 'merely' out of phase with, SDP.

A major feature of most LDP is their real need for a long light period, rather than just a short dark period. That is, in most LDP, brief night-break interruptions are *not* effective in promoting flowering – either a long light period or a long night-break is required. This need for a substantial light period has led to the term **light-dominant plant** (which, since a few species can be induced to flower by brief night-break is not exactly analogous to the traditional term, LDP).

Further characteristics of the light periods required by light-dominant plants are:

(1) a specific mixture of red and far-red irradiation is more effective than either spectral region alone (in many cases, the most effective ratio of R : Fr changes over the course of the light period);

(2) action maxima for greatest effectiveness of light extension of a suboptimal photoperiod, are around λs 710 nm to 720 nm;

(3) there are usually strong quantitative effects of irradiance.

Thus, the light requirements of light-dominant plants show many of the characteristics of the HIR. Whether this means that a different photoperiodic mechanism is involved in LDP is not yet known.

[9] See Deitzer (1984), Vince-Prue (1983a).

8.4.4 Nature of the photoperiodic signal[10]

Two types of change occur in the light environment at the beginning and end of the day – change in light quality, and change in quantity. The actual feature that phytochrome detects as the end-of-day signal has been the subject of some debate.

Quality
At dusk, there is a change in the R : Fr ratio and in the relative amounts of blue light in the spectrum (Ch. 3). However, at high latitudes, such changes are relatively small and occur gradually. Furthermore, under vegetation canopies themselves, the R : Fr ratio is lowered (in some situations, to much lower values than those which characterize the onset of twilight – see Table 3.3). Therefore, it is considered that spectral change itself is not reliable enough to act as the end-of-day signal. This view is reinforced by the results of actual experiments where the end-of-day R : Fr ratio is varied: although the extent of flowering is often affected, the timing of photoperiodic sensitivity is invariably unaltered.

Quantity
A change in the irradiance during the light period, however, does have photoperiodic effects. Therefore, it is thought that the *fall in irradiance* at dusk is the signal that initiates photoperiodic dark-time measurement. An important qualification, though, is that dusk need only deepen to a 'physiological darkness', rather than to absolute darkness (i.e. photoperiodic time measurement is initiated when there is still a low level of light present). There is a certain amount of species variation in actual light sensitivity, but in general, light levels lower than 50–100 mW m^{-2} are photoperiodically equivalent to darkness. If 'darkness' is thus considered as a particular low threshold of light, the timing of daylength by the fall in irradiance at dusk can, in fact, be quite precise: actual measurements have shown that under natural conditions, the point of 'darkness' (as a particular low irradiance) falls within the same 5–12-minute period over the course of a few days.

How phytochrome actually detects (or responds to) a light-off signal remains a major puzzle. At the point of light-off, phytochrome is in the form of P_{fr}, and recent work has indicated that in light-grown plants, P_{fr} remains stable for many hours in subsequent darkness – that is, the low levels of phytochrome in green plants do not show the physiological dark reactions exhibited by etiolated seedlings. (Also, recall that photochemical transformation of P_{fr} prior to darkness by end-of-day far-red irradiation does not affect the length of the critical

[10] See Smith & Morgan (1983), Vince-Prue (1983b).

dark period.) Therefore, it is not clear exactly what happens to P_{fr} at the point of light-off, to allow dark time measurement to begin. Suggested mechanisms (Vince-Prue 1983b) by which phytochrome could 'sense' the cessation of light include:

(1) the existence of separate pools of phytochrome, of which a labile pool is involved in time measurement (dark-degradation would start at light-off);

(2) the involvement of continual photocycling between the P_r and P_{fr} forms, or the involvement of the photo-intermediates, in the biological activity of phytochrome (phytochrome activity would thus cease at light-off);

(3) the involvement only of 'new' P_{fr} in biological activity (this suggestion derives from phytochrome-mediated responses like the control of chloroplast movement in *Mesotaenium*, where repeated further exposures to red light are necessary, even though P_{fr} is already detectable – it is thought that P_{fr} 'ages' and becomes inactive).

In relation to our present ideas of its mode of action, phytochrome does not seem suited to the detection of light-cessation (a more appropriate type of photoreceptor would seem to be one which initiates reaction from an excited state – activity would stop upon transfer to darkness). However, the fact that phytochrome does seem to be the detector of the light–dark transition merely highlights our considerable lack of basic knowledge concerning phytochrome action.

8.5 SUMMARY

(1) Many of the daily activities of a plant are controlled by an endogenous circadian rhythm. Light interacts with circadian rhythms in various ways, which include phase-shift responses and rhythm entrainment. Various types of photoreceptors interact with circadian rhythms by acting as coupling agents which transmit information about the external light cycle into the endogenous timing mechanism; phytochrome and, in some cases, blue-light receptors, interact with rhythms in higher plants.

(2) The circadian timing mechanism may involve some property of the cell membrane, perhaps control of certain ion fluxes. A photoreceptor may exert its effects on advance or delay of circadian timing through changing the properties of this membrane.

(3) Seasonal control of development by photoperiodism enables a plant to be in synchrony with climatic conditions and with other organisms. There are various photoperiodic classes of plant, distinguished by their flowering responses to changing day-length. There is great variation in photoperiodic sensitivity and some species have ecotypes which respond differently to daylength.

(4) The photoperiodic signal is probably the fall in light quantity at dusk; this fall in irradiance is perceived by phytochrome in the leaves. The photoperiodic signal entrains a circadian rhythm in light sensitivity: a positive photoperiodic response occurs when the dawn signal coincides with the appropriate stage of this rhythm.

(5) While LDP also exhibit a rhythm in light sensitivity, most LDP have a requirement for an extensive light period.

(6) Flowering in many species is, or becomes, independent of photo-period. Other major controls of flowering are age (juvenility), temperature (vernalization) and nutrition.

FURTHER READING

Palmer, J. D. 1976. *An introduction to biological rhythms*. London: Academic Press.

Vince-Prue, D. 1975. *Photoperiodism in plants*. Maidenhead: McGraw-Hill.

Vince-Prue, D., B. Thomas & K. E. Cockshull (eds.) 1984. *Light and the flowering process*. London: Academic Press.

Glossary

absorption coefficient (symbol ε)   An expression of the capacity of a molecule to absorb light; it has a characteristic value for a substance in relation to its molecular environment (e.g. type of solvent), and the particular wavelengths of light involved; the older term for this parameter is *extinction coefficient*.

absorption filter   A type of light filter that transmits certain waveband regions, but reflects or absorbs others.

absorption spectrum   The wavelengths of radiant energy specifically absorbed by a substance; usually presented as a plot of the amount of light absorbed, against wavelength; the wavelengths that are maximally absorbed are often specified as λ_{max} (lambda).

actinic light   Light that produces a biological response or a biochemical change, i.e. a 'light treatment', as opposed to light that is used for assay or measurement purposes.

action spectrum   A plot against wavelength of the relative effectiveness of different wavelengths of light in bringing about a particular biological response.

autotroph   An organism that synthesizes its own food from inorganic molecules; photosynthetic organisms are photo-autotrophs.

Avogadro's number   The number of molecules in the gram-molecular weight of a substance; equal to 6.022×10^{23} mol^{-1}; hence, in photobiology, a 'mole of light' contains 6.022×10^{23} photons.

blue light photoreceptors   The receptors responsible for light absorption in specific responses to blue and near-ultraviolet radiation; candidates for the light-absorbing part of the receptor include carotenoid and flavin types of molecules.

calmodulin   A particular protein which reversibly binds calcium, and whose activity results in the regulation of the activities of other proteins (e.g. enzymes).

chromophore   A pigment molecule that is the part of a photoreceptor that absorbs light.

circadian rhythm   A regular fluctuation in a biological activity, based on an endogenous timing mechanism with a period or cycle length of approximately 24 hours; such endogenous daily rhythms are entrained by some exogenous factor, often light.

critical daylength   The length of light period that is critical for the expression of a photoperiodic response; short-day plants do not flower if the daylength is *longer* than their critical daylength, and long-day plants do not flower in light periods *shorter* than their critical period; (sometimes the alternative term 'critical night length' is used – short-day plants do not flower if darkness is shorter than their critical night length).

cryptochrome   Generic term for photoreceptors involved in specific responses to blue–ultraviolet light (becoming less used since there is probably more than one type of blue-light receptor).

dichroism   Differential absorption of light which is plane polarized in different directions – that is, the chemical make-up and orientation of the absorbing molecule is such that it preferentially absorbs light vibrating in a particular plane.

dormancy   Resting state of buds, seeds and spores that fail to grow when provided with (seemingly) optimal conditions; dormant organs need some special treatment to 'break' the dormancy and initiate growth; seed which require a light treatment in order to germinate are said to be photodormant.

einstein   Older term for a mole of photons, i.e. a number of photons equal to Avogadro's number of 6.022×10^{23}.

emission spectrum   Characteristic wavelengths of radiant energy emitted from an excited atom or molecule; also describes the characteristic wavelengths that are emitted from a source of artificial light.

energy fluence rate   See **fluence rate**.

energy spectral distribution   See **spectral distribution**.

escape time   Relates to the action of phytochrome and describes the length of time between the end of a red light induction treatment for a response, and the subsequent loss of reversibility by far-red irradiation (it describes the length of time after which the action of phytochrome cannot be reversed).

etiolation   Describes the state of a plant that has been grown in conditions of low or no light; there can be various degrees of etiolation depending on the actual amount of light received; in the complete absence of light, most higher plants show long stems, undeveloped leaves, and no chlorophyll or other pigmentation.

excited state   A high energy, unstable state of an atom or molecule; it can be brought about by various means including thermal effects, absorption of radiant energy and electric discharge.

extinction coefficient   See **absorption coefficient**.

eyespot   The photosensitive area at the base of the flagellum in certain phototactic lower plants; it consists of a screening pigment and a photoreceptor, and functions to regulate flagellar activity in relation to the direction of light.

fluence   The amount of light applied per unit area; expressed as energy fluence ($J\ m^{-2}$) or as photon fluence ($mol\ m^{-2}$); sometimes equated with light 'dose' (although this is technically wrong because of the distinction between light applied and dose actually absorbed).

fluence rate   Expression of the 'flow rate' of light, i.e. the amount of light per unit area per unit time; it can be expressed as **energy fluence rate** (energy per area per time: J m^{-2} s^{-1} or W m^{-2}), or as **photon fluence rate** (number of photons per area per time: mol m^{-2} s^{-1}); (older terms are energy flux density and photon flux density); conversion between energy fluence rate and photon fluence rate is:

$$\text{mol m}^{-2}\text{ s}^{-1} = \text{W m}^{-2} \times \frac{\lambda(\text{metres})}{0.1196}$$

frequency (symbol v, nu) Description of radiant energy in terms of the number of wave crests (peaks in energy) passing a stationary point in one second (red light has a frequency of 5×10^{14} cycles s^{-1}).

ground state The lowest stable energy level of an atom or molecule.

heliotropism The term used originally to describe phototropism; used nowadays to describe the phenomenon of 'Sun tracking' by organs such as leaves and flowers.

heterotroph An organism which cannot synthesize its own food, and which depends upon previously synthesized organic foodstuffs.

high irradiance reaction (HIR) The photosystem operating in photomorphogenic responses characterized by requirements for high fluence rates and long exposure times, and with action maxima in one or more of the red, far-red or blue regions of the spectrum and no red–far-red reversibility.

interference filter A type of light filter that uses optical phenomena of interference to produce narrow wavebands of radiation; interference filters are composed of two semi-transparent layers whose precise spacing causes multiple reflection and constructive reinforcement of the required wavelengths, but destructive interference of other wavelengths.

irradiance Radiant energy received on a unit surface per unit time; energy fluence rate (W m^{-2} or J s^{-1} m^{-2}).

light The visible portion of the electromagnetic spectrum, λ 380 nm to λ 780 nm; the term also generally includes the near-ultraviolet and near-infra-red ends of the visible region; speed of light in a vacuum = 2.998×10^8 m s^{-1}.

low energy response (LER) Inductive (triggered) photomorphogenic responses mediated by phytochrome; characterised by saturation at low fluences (10–100 J m^{-2}), induction by brief irradiation treatments, and showing red–far-red reversibility.

mole Gram-molecular weight of a substance – that is, the weight of the substance in grams, numerically equal to its molecular weight; every mole of any substance contains Avogadro's number of molecules.

nanometre 10^{-9} metres; equivalent to the older term millimicron (mµ).

nastic movements (sleep movements) Reversible movements of organs such as leaves and flowers; often triggered by an external stimulus, but the direction of movement is determined by the anatomy of the organ itself, and is usually based upon turgor changes in the motor cells of a pulvinus; many daily nastic movements also show a strong circadian rhythm (the suffix '-nasty' indicates that a movement is determined by endogenous rather than exogenous factors).

neutral density filter A type of light filter that changes the quantity of light rather than the quality of light (i.e. it reduces the amount of light uniformly across the spectrum).

nyctinasty Daily nastic movements, especially of leaves which open during the day and fold at night; the movements show a strong circadian rhythm, which is reinforced by the external cycle of light and darkness (see also **nastic movements**, **photonasty**).

photoblastic Refers to seeds whose germination is affected by light; germination is stimulated by light in positively photoblastic seeds, and inhibited by light in negatively photoblastic seeds.

photochemical reaction A reaction which occurs directly in response to a substrate becoming excited to a higher energy level by the absorption of radiant energy; in the absence of any chain reaction effects, the number of chemically changed molecules is directly equal to the number of photons absorbed (see also **quantum yield**); truly photochemical reactions are largely independent of temperature.

photochromicity Light-reversible changes in the absorption properties of a pigment.

photoconversion cross section (symbol σ) An expression of the probability that a photon will bring about a photoreaction; it is related to the two parameters of quantum yield of the reaction and extinction coefficient of the reacting molecule.

photoequilibrium (of phytochrome; symbol ϕ, phi) Describes the relative amounts of the two different forms of phytochrome present in a particular light environment; it derives solely from the actions of light and is calculated from the expression $\phi = k_1/(k_1 + k_2)$, where k_1 and k_2 are the rate constants for the photoconversions of P_r and P_{fr}; k_1 and k_2 are themselves calculated from incident photon flux, extinction coefficients and quantum yields (see also **photostationary state**).

photokinesis Refers to the rate of movement in a motile lower organism being affected by light.

photomorphogenesis The non-photosynthetic influence of light on germination, growth, development and reproduction; some usages also include other regulatory aspects of light on plant form and orientation, (e.g. such non-developmental processes as phototropism, photonasty etc.).

photon A quantum (or discrete particle) of radiant energy; used particularly to describe quanta in the visible region.

photonasty Daily nastic movements, especially of flowers which open at one part of the daily cycle and close at others; the term is also used to describe the protective 'light-avoiding' movements shown by the leaves of certain species in strong sunlight (see also **nastic movements, nyctinasty**).

photon spectral distribution See **spectral distribution**.

photoperiodism The influence of daylength (or the regular periodicity of light and darkness) on growth and development.

photophil phase A term used in relation to one of the concepts presently suggested to account for photoperiodism; it refers to the circadian phase of growth during which light is stimulatory or beneficial to development (see also **skotophil phase**).

photoreceptor A molecule or biological structure that absorbs light and directly initiates a chain of metabolic and physiological events that result in some biological response; a photoreceptor is thus the component that is responsible for the transduction of radiant energy into some other form of chemical or electrical energy that is biologically usable.

photostationary state (of phytochrome) Describes the ratio of $P_{fr} : P_{total}$ in a particular light environment; its value is determined not only by the actions of light, but also by the effects of certain 'dark' reactions, including synthesis and degradation, and it is derived from spectrophotometric measurements of phytochrome in etiolated material in the appropriate light environments (see also **photoequilibrium**).

photosystem The process or chain of events by which light brings about some biological response; a photosystem consists of various components or subprocesses: components involved in light absorption (the photoreceptor); components involved in the transduction and amplification of radiant energy into biochemical energy (the primary reaction of the photoreceptor and associated events in other molecules); components involved in the direct utilization of this biochemical energy.

phototaxis The movement of an organism in relation to the direction of light, positively (towards the light) or negatively (away from the light).

phototropism A directional growth response to a directional light stimulus (see also **tropism, heliotropism**).

phytochrome The photoreceptor responsible for the photomorphogenic responses of plants to red and far-red light; it consists of a protein and its attached chromophore (a linear tetrapyrrole chemically related to the bile pigments); phytochrome is photochromic, the two isomeric forms being interconvertible by red and far-red irradiation.

pigment A coloured substance; a molecule that absorbs certain wavelengths of visible light and reflects others; a pigment molecule comprises an extensive π orbital system of conjugated double (C=C) and single (C–C) bonds.

Planck's constant A universal constant which relates the energy of a photon to the frequency of the radiant energy; its dimensions are energy and time and it is equal to 6.6×10^{-34} J s.

plastid The generic term for a class of organelles that is unique to plants; they are characterized by a double bounding membrane and an internal membrane system; different types are distinguished by their pigment content and internal structure, and include etioplasts, chloroplasts, amyloplasts, etc.

polarised light Light normally consists of electric and magnetic vibrations taking place in all possible planes at right angles to the direction of the light path; in plane polarized light, the vibrations are restricted to particular planes; the extent of polarization varies, and it can be brought about by passing light through a special filter or as a consequence of reflection.

polarotropism A directional response of an organ or organelle to the plane of polarization of light; results from the differential absorption of light by a dichroically oriented photoreceptor.

prolamellar body The characteristic structure within the plastids of dark-grown tissues (i.e. within the etioplasts); it is quasi-crystalline in appearance and consists mainly of membrane lipid materials.

protonema (plural, protonemata) The initial filamentous stage in the development of the haploid gametophyte of many liverworts, mosses and ferns.

pulvinus A swollen region, usually at the base of a leaf or flower part, which is responsible for the nastic movements of the organ; the pulvinus contains special cells which swell and contract and thus bring about movement of the organ.

pyrrole ring A five-membered molecular ring structure that is made up of four carbon atoms and one nitrogen atom; it is a characteristic structural feature of haem-like molecules (e.g. chlorophyll, phytochrome, cytochromes).

quantum A single indivisible pulse or packet of radiant energy.

quantum sensor An instrument for measuring amount of light as number of photons; it consists of a photovoltaic cell through which photons generate an electric current; a series of filters 'quantum correct' the energies to give a direct relationship between number of photons and voltage; commonly available instruments are only sensitive to λs 400–700 nm.

quantum yield (symbol Φ phi) An expression that relates the number of molecules changed in a photochemical reaction (M), to the number of photons absorbed (P): $\Phi = \frac{M}{P}$; a truly photochemical reaction has a quantum yield of 1.0; in photomorphogenic responses, the 'quantum yield' is usually much greater than 1.0, due to biological amplification mechanisms in the photosystem subsequent to the actual photoreaction.

radiometer An instrument for measuring amounts of radiant energy; it consists of a thermopile across which radiant energy generates a temperature-induced electromotive force.

red : far-red ratio (symbol ζ) The ratio of the photon fluence rate between λs 655–665 nm and λs 725–735 nm; it is used to describe the spectral distribution in an environment in terms relevant to the operation of phytochrome.

scattering (of light) When light travels through a medium, scattering of the beam takes place; there are two general types of scattering: random *reflection* occurs when the scattering particles are larger than the light wavelengths; *diffraction* occurs when the scattering molecules are smaller than the wavelengths; the extent of diffraction is inversely proportional to the wavelength (i.e. blue light is scattered more than red light).

skotophil phase A term used in relation to one of the concepts presently suggested to account for photoperiodism; it refers to the circadian phase of growth which requires darkness – that is, the phase during which exposure to light is inhibitory or detrimental to development (see also **photophil phase**).

sleep movements See **nastic movements**.

solarimeter See **radiometer**.

spectral distribution A description of the composition of the light in a particular situation; it can be expressed as *energy spectral distribution* – that is, energy fluence rate per wavelength ($J s^{-1} \lambda^{-1} m^{-2}$ or $W \lambda^{-1} m^{-2}$); or it can be expressed as *photon spectral distribution* – that is, photon fluence rate per wavelength ($mol \lambda^{-1} m^{-2} s^{-1}$); it is measured by scanning the spectrum with a spectroradiometer.

spectroradiometer An instrument for analysing the spectral distribution in an environment; a monochromator continually scans the spectrum and the amounts of light in the different waveband regions are measured.

thermonasty Nastic movements regulated by changes in temperature; exemplified by flower opening and closing in crocus, which occurs through regular changes in the growth rates on the inner and outer sides of the petals in response to changes in the day and night-time temperatures.

thermoperiodism The beneficial effects on plant growth of alternating higher day-time and lower night-time temperatures.

thigmonasty Nastic movements triggered by contact stimulation; exemplified by leaflet closure in *Mimosa pudica* in response to touch.

tropism A directional growth response to a directional environmental stimulus; the direction of response can be *positive* (towards), *negative* (away from), *diatropic* (at right angles to) or *plagiotropic* (at some other angle to) the direction of stimulus.

vernalization The enhancement of subsequent flowering responses by low temperature ($0°–10°C$) treatment of imbibed seeds or young seedlings.

wavelength (of light) (symbol λ lambda) The description of electromagnetic radiation in terms of the distance between two consecutive crests (or energy peaks); the energy in a photon is inversely proportional to its wavelength (i.e. short wavelength radiation is more energetic than long wavelengths).

wavenumber (of light) (symbol ṽ) Description of electromagnetic radiation in terms of the number of wave crests (or energy peaks) present in one centimetre of radiation (i.e. another way of expressing frequency).

References

Bickford, E. D. & S. Dunn 1972. *Lighting for plant growth*. Ohio: Kent State University Press.

Bjorn, L. O. 1976. *Light and life*. London: Hodder & Stoughton.

Bjorn, L. O. 1980. Blue light effects on plastid development in higher plants. In *The blue light syndrome*, H. Senger (ed.), 455–65. Berlin: Springer.

Blaauw, O. H. & G. Blaauw-Jansen 1970. The phototropic responses of *Avena* coleoptiles. *Acta Bot. Neerl.* **19**, 755–63.

Black, M. & A. J. VLitos 1972. Possible inter-relationships of phytochrome and plant hormones. In *Phytochrome*, K. Mitrakos & W. Shropshire (eds), 517–52. London & New York: Academic Press.

Briggs, W. R. & M. Iino 1983. Blue light absorbing photoreceptors in plants. *Phil Trans R. Soc.* **B303**, 347–57.

Buchanan, B. B. 1980. Role of light in the regulation of chloroplast enzymes. *Ann. Rev. Plant Physiol.* **31**, 341–74.

Castelfranco, P. A. & S. I. Beale 1983. Chlorophyll biosynthesis: recent advances and areas of current interest. *Ann. Rev. Plant Physiol.* **34**, 241–78.

Clayton, R. K. 1970. *Light and living matter: a guide to the study of photobiology*. Volume 1: The physical part. New York: McGraw-Hill.

Cockshull, K. E. 1984. The photoperiodic induction of flowering in short-day plants. In *Light and the flowering process*, D. Vince-Prue, B. Thomas & K. E. Cockshull (eds), 33–50. London: Academic Press.

Cosgrove, D. J. 1981. Rapid suppression of growth by blue light. Occurrence, time course and general characteristics. *Plant Physiol.* **67**, 584–90.

Cosgrove, D. J. 1983. Photocontrol of extension growth: a biophysical approach. *Phil Trans R. Soc.* **B303**, 453–65.

Cosgrove, D. J. 1985. Kinetic separation of phototropism from blue-light inhibition of stem growth. *Photochem. Photobiol.* **42**, 745–51.

Darwin, C. 1880. *The power of movement in plants*. London: John Murray.

DeGreef, J. A. & H. Fredericq 1983. Photomorphogenesis and hormones. In *Photomorphogenesis*, Encycl. Plant Physiol. NS 16A, W. Shropshire & H. Mohr (eds), 401–27. Berlin: Springer.

Deitzer, G. F. 1984. Photoperiodic induction in long-day plants. In *Light and the flowering process*, D. Vince-Prue, B. Thomas & K. E. Cockshull (eds), 51–64. Londor: Academic Press.

Dennison, D. S. 1979. Phototropism. In *Physiology of movements*, Encycl. Plant Physiol. NS 7, W. Haupt & M. E. Feinleib (eds) 506–66. Berlin: Springer.

Dennison, D. S. 1984. Phototropism. In *Advanced plant physiology*, M. B. Wilkins (ed.), 149–62. London: Pitman.

Ellis, R. J. 1984. Kinetics and fluence-response relationships of phototropism in the dicot *Fagopyrum esculentum* (buckwheat). *Plant Cell Physiol.* **25**, 1513–20.

195

Firn, R. D. & J. Digby 1980. The establishment of tropic curvatures in plants. *Ann. Rev. Plant Physiol.* **31**, 131–48.

Foster, K. W., J. Saranak, N. Patel, G. Zorilli, M. Okabe, T. Kline & K. Nakanishi 1984. A rhodopsin is the functional photoreceptor for phototaxis in the unicellular eukaryote *Chlamydomonas*. *Nature* **311**, 756–9.

Frankland, B. 1981. Germination in the shade. In *Plants and the daylight spectrum*, H. Smith (ed.), 187–204. London: Academic Press.

Frankland, B. & R. Taylorson 1983. Light control of seed germination. In *Photomorphogenesis*, Encycl. Plant Physiol. NS 16A, W. Shropshire & H. Mohr (eds), 428–56. Berlin: Springer.

Gaba, V. & M. Black 1983. The control of cell growth by light. In *Photomorphogenesis*, Encycl. Plant Physiol. NS 16A, W. Shropshire & H. Mohr (eds), 358–400. Berlin: Springer.

Galland, P. 1983. Action spectra of photogeotropic equilibrium in *Phycomyces* wild type and three behavioural mutants. *Photochem. Photobiol.* **37**, 221–8.

Galston, A. W. 1983. Leaflet movements in *Samanea*. In *The biology of photoreception*, SEB Symposium No. 36, D. Cosens & D. Vince-Prue (eds), 541–59. Cambridge: Cambridge University Press.

Giese, A. C. 1978. *Living with our sun's ultraviolet rays*. New York: Plenum Press.

Gordon, D. C., I. R. MacDonald, J. W. Hart & A. R. Berg 1984. Image analysis of geo-induced inhibition, compression and promotion of growth in an inverted *Helianthus annuus* seedling. *Plant Physiol.* **76**, 589–94.

Grime, J. P. 1981. Plant strategies in shade. In *Plants and the daylight spectrum*, H. Smith (ed.), 159–86. London: Academic Press.

Hahlbrock, K. & H. Grisebach 1979. Enzymic controls in the biosynthesis of lignin and flavonoids. *Ann. Rev. Plant Physiol.* **30**, 105–30.

Halevy, A. H. 1984. Light and autonomous induction. In *Light and the flowering process*, D. Vince-Prue, B. Thomas & K. E. Cockshull (eds) 65–74. London: Academic Press.

Hart, J. W., D. C. Gordon and I. R. MacDonald 1982. Analysis of growth during phototropic curvature of cress hypocotyls. *Plant, Cell & Environ.* **5**, 361–6.

Haupt, W. 1983. The perception of light direction and orientation responses in chloroplasts. In *The biology of photoreception*, SEB Symposium No. 36, D. Cosens & D. Vince-Prue (eds), 423–42. Cambridge: Cambridge University Press.

Hayward, P. M. 1984. Phytochrome parameters. In *Techniques in photomorphogenesis*, H. Smith & M. G. Holmes (eds), 158–73. London: Academic Press.

Holmgren, A. 1985. Thioredoxin. *Ann. Rev. Biochem.* **54**, 237–71.

Holmes, M. G. 1981. Spectral distribution of radiation within plant canopies. In *Plants and the daylight spectrum*, H. Smith (ed.), 147–58. London: Academic Press.

Holmes, M. G. 1983. Perception of shade. *Phil Trans R. Soc.* **B303**, 503–21.

Holmes, M. G. 1984a. Light sources. In *Techniques in photomorphogenesis*, H. Smith & M. G. Holmes (eds), 43–80. London: Academic Press.

Holmes, M. G. 1984b. Radiation measurement. In *Techniques in photomorphogenesis*, H. Smith & M. G. Holmes (eds), 81–108. London: Academic Press.

Iino, M. & E. Schäfer 1984. Phototropic response of the stage I *Phycomyces*

sporangiophore to a pulse of blue light. *Proc. Nat. Acad. Sci. (US)* **81**, 7103–7.
Iino, M., W. R. Briggs & E. Schäfer 1984. Phytochrome mediated phototropism in maize seedling shoots. *Planta* **160**, 41–51.

Jabben, M. & M. G. Holmes 1983. Phytochrome in light-grown plants. In *Photomorphogenesis*, Encycl. Plant Physiol. NS 16B, W. Shropshire & H. Mohr (eds), 704–22. Berlin: Springer.
Jenkins, G. I., M. R. Hartley & J. Bennett 1983. Photoregulation of chloroplast development: transcriptional, translational and post-translational controls. *Phil Trans R. Soc.* **B303**, 419–29.

Kasemir, H. 1983. Light control of chlorophyll accumulation higher plants. In *Photomorphogenesis*, Encycl. Plant Physiol. NS 16B, W. Shropshire & H. Mohr (eds), 662–86. Berlin: Springer.
Kendrick, R. E. 1983. The physiology of phytochrome action. In *The biology of photoreception*, SEB Symposium No. 36, D. Cosens & D. Vince-Prue (eds), 275–304. Cambridge: Cambridge University Press.
King, R. W. 1984. Light and photoperiodic timing. In *Light and the flowering process*, D. Vince-Prue, B. Thomas & K. E. Cockshull (eds), 91–105. London: Academic Press.
Koorneef, M., E. Rolff & C. J. P. Spruit 1980. Genetic control of light-inhibited hypocotyl elongation. *Z. Pflanzenphysiol.* **100**, 147–60.

Lagarias, J. C. 1985. Progress in the molecular analysis of phytochrome. *Photochem. Photobiol.* **42**, 811–20.
Lamb, C. J. & M. A. Lawton 1983. Photocontrol of gene expression. In *Photomorphogenesis*, Encycl. Plant Physiol. NS 16A, W. Shropshire & H. Mohr (eds), 213–57. Berlin: Springer.
Lichtenthaler, H. K., C. Buschman & U. Rahmsdorf 1980. The importance of blue light for the development of sun-type chloroplasts. In *The blue light syndrome*, H. Senger (ed.), 485–94. Berlin: Springer.

MacDonald, I. R., D. C. Gordon, J. W. Hart & P. Maher 1983. The positive hook: the role of gravity in the formation and maintenance of the apical hook. *Planta* 158, 76–81.
Macleod, K., J. Digby & R. D. Firn 1985. Evidence inconsistent with the Blaauw model of phototropism. *J. Exp. Bot.* **36**, 312–19.
Mancinelli, A. L. 1980. The photoreceptors of the high irradiance responses of plant photomorphogenesis. *Photochem. Photobiol.* **32**, 853–7.
Mancinelli, A. L. 1983. The photoregulation of anthocyanin synthesis. In *Photomorphogenesis*, Encycl. Plant Physiol. NS 16B, W. Shropshire & H. Mohr (eds), 640–61. Berlin: Springer.
Mandoli, D. F. & W. R. Briggs 1981. Phytochrome control of two low irradiance responses in etiolated oat seedlings. *Plant Physiol.* **67**, 733–9.
Mansfield, T. A. & P. J. Snaith 1984. Circadian rhythms. In *Advanced plant physiology*, M. B. Wilkins (ed.), 201–18. London: Pitman.
Mohr, H. 1972. *Lectures in photomorphogenesis*. Berlin: Springer.
Mohr, H. 1983. Pattern specification and realisation in photomorphogenesis. In *Photomorphogenesis*, Encycl. Plant Physiol. NS 16A, W. Shropshire & H. Mohr (eds), 336–57. Berlin: Springer.
Mohr, H. 1984. Criteria for photoreceptor involvement. In *Techniques in*

photomorphogenesis, H. Smith & M. G. Holmes (eds), 13–42. London: Academic Press.

Morgan, D. C. 1981. Shadelight quality effects on plant growth. In *Plants and the daylight spectrum*, H. Smith (ed.), 205–22. London: Academic Press.

Morgan, D. C., R. Child & H. Smith 1981. Absence of fluence rate dependency of phytochrome modulation of stem extension in light grown *Sinapis alba*. *Planta* **151**, 497–8.

Napp-Zinn, K. 1984. Light and vernalisation. In *Light and the flowering process*, D. Vince-Prue, B. Thomas & K. E. Cockshull (eds), 75–88. London: Academic Press.

Newbury, H. J. 1983. Photocontrol of enzyme activation and inactivation. *Phil Trans R. Soc.* **B303**, 433–40.

Njus, D., F. M. Sulzman & J. W. Hastings 1974. Membrane model for the circadian clock. *Nature* **248**, 116–20.

Palmer, J. D. 1976. *An introduction to biological rhythms.* London: Academic Press.

Poff, K. L. 1983. Perception of a unilateral light stimulus. *Phil Trans R. Soc.* **B303**, 479–87.

Possingham, J. V. 1980. Plastid replication and development in the life cycle of higher plants. *Ann. Rev. Plant Physiol.* **31**, 113–29.

Pratt, L. H. 1982. Phytochrome: the protein moiety. *Ann. Rev. Plant Physiol.* **33**, 557–82.

Pratt, L. H. 1983. Assay of photomorphogenic receptors. In *Photomorphogenesis*, Encycl. Plant Physiol. NS 16A, W. Shropshire & H. Mohr (eds), 152–77. Berlin: Springer.

Presti, D. E. 1983. The biology of carotenes and flavins. In *The biology of photoreception*, SEB Symposium No. 36, D. Cosens & D. Vince-Prue (eds), 133–80. Cambridge: Cambridge University Press.

Presti, D. E. & M. Delbruck 1978. Photoreceptors for biosynthesis, energy storage and vision. *Plant, Cell & Environ.* **1**, 81–100.

Quail, P. H. 1983. Rapid action of phytochrome in photomorphogenesis. In *Photomorphogenesis*, Encycl. Plant Physiol. NS 16A, W. Shropshire & H. Mohr (eds), 178–212. Berlin: Springer.

Quail, P. H., J. T. Colbert, H. P. Hershey & R. D. Vierstra 1983. Phytochrome: molecular properties and biogenesis. *Phil Trans R. Soc.* **B303**, 387–402.

Racusen, R. H. & A. W. Galston 1983. Developmental significance of light-mediated electrical responses in plant tissue. In *Photomorphogenesis*, Encycl. Plant Physiol. NS 16B, W. Shropshire & H. Mohr (eds), 687–703. Berlin: Springer.

Raven, J. A. 1983. Do plant photoreceptors act at the membrane level? *Phil Trans R. Soc.* **B303**, 403–18.

Rich, T. C. G., G. C. Whitelam & H. Smith 1985. Phototropism and axis extension in light-grown mustard (*Sinapis alba* L.) seedlings. *Photochem. Photobiol.* **42**, 789–92.

Roux, S. J. 1983. A possible role for Ca^{2+} in mediating phytochrome responses. In *The biology of photoreception*, SEB Symposium No. 36, D. Cosens & D. Vince-Prue (eds), 561–80. Cambridge: Cambridge University Press.

Rudiger, W. & H. Scheer 1983. Chromophores in photomorphogenesis. In

Photomorphogenesis, Encycl. Plant Physiol. NS 16A, W. Shropshire & H. Mohr (eds), 119–51. Berlin: Springer.

Salisbury, F. B. & C. W. Ross 1978. *Plant physiology*, 2nd edn. Belmont, California: Wadsworth.

Schaer, J. A., D. F. Mandoli & W. R. Briggs 1983. Phytochrome-mediated cellular photomorphogenesis. *Plant Physiol.* **72**, 706–12.

Schäfer, E. & L. Fukshansky 1984. Action spectroscopy. In *Techniques in photomorphogenesis*, H. Smith & M. G. Holmes (eds), 109–29. London: Academic Press.

Schäfer, E. & W. Haupt 1983. Blue light effects in phytochrome-mediated responses. In *Photomorphogenesis*, Encycl. Plant Physiol. NS 16B, W. Shropshire & H. Mohr (eds), 723–44. Berlin: Springer.

Schmidt, W. 1983. The physiology of blue light systems. In *The biology of photoreception*, SEB Symposium No. 36, D. Cosens & D. Vince-Prue (eds), 305–30. Cambridge: Cambridge University Press.

Schopfer, P. 1977. Phytochrome control of enzymes. *Ann. Rev. Plant Physiol.* **77**, 223–52.

Schopfer, P. 1984. Photomorphogenesis. In *Advanced plant physiology*, M. B. Wilkins (ed.), 380–407. London: Pitman.

Schopfer, P. & K. Apel 1983. Intracellular photomorphogenesis. In *Photomorphogenesis*, Encycl. Plant Physiol. NS 16A, W. Shropshire & H. Mohr (eds), 258–88. Berlin: Springer.

Schwartz, A. & D. Koller 1980. Role of the cotyledons in the phototropic response of *Lavatera cretica* seedlings. *Plant Physiol.* **66**, 82–7.

Seliger, H. H. & W. D. McElroy 1965. *Light: physical and biological action*, ch. 1, 2. New York: Academic Press.

Senger, H. 1982. The effect of blue light on plants and micro-organisms. *Photochem. Photobiol.* **35**, 911-20.

Senger, H. & W. R. Briggs 1981. The blue light receptor(s): primary actions and subsequent metabolic changes. In *Photochem. Photobiol. Rev.* vol. 6, K. C. Smith (ed.), 1–38. New York: Plenum.

Smith, H. 1982. Light quality, photoperception and plant strategy. *Ann. Rev. Plant Physiol.* **33**, 481–518.

Smith, H. 1983a. Is P_{fr} the active form of phytochrome?: *Phil Trans R. Soc.* **B303**, 443–52.

Smith, H. 1983b. The natural radiation environment: limitations on the biology of photoreceptors. Phytochrome as a case study. In *The biology of 'photoreception*, SEB Symposium No. 36, D. Cosens & D. Vince-Prue (eds), 1–18. Cambridge: Cambridge University Press.

Smith, H. 1984. Plants that track the sun. *Nature* **308**, 774.

Smith, H. & D. C. Morgan 1983. The function of phytochrome in nature. In *Photomorphogenesis*, Encycl. Plant Physiol. NS 16B, W. Shropshire & H. Mohr (eds), 491–517. Berlin: Springer.

Smith, W. O. 1983. Phytochrome as a molecule. In *Photomorphogenesis*, Encycl. Plant Physiol. NS 16A, W. Shropshire & H. Mohr (eds), 96–118. Berlin: Springer.

Song, P.-S. 1984. Phytochrome. In *Advanced plant physiology*, M. B. Wilkins (ed.), 354–79. London: Pitman.

Spence, D. H. N. 1981. Light quality and plant responses underwater. In *Plants and the daylight spectrum*, H. Smith (ed.), 245–76. London: Academic Press.

Sweeney, B. M. 1977. Chronobiology (circadian rhythms). In *The science of photobiology*, K. C. SMith (ed.), 209–26. New York: Plenum.

Sweeney, B. M. 1979. Endogenous rhythms in the movement of plants. In *Physiology of movements*, Encycl. Plant Physiol. NS 7, W. Haupt & M. E. Feinleib (eds), 71–94. Berlin: Springer.

Taiz, L. 1984. Plant cell expansion: regulation of cell wall mechanical properties. *Ann. Rev. Plant Physiol.* **35**, 585–657.

Thimann, K. V. 1977. *Hormones and the whole life of plants*. Boston, Mass.: University of Massachusetts Press.

Thomas, B. 1981. Specific effects of blue light on plant growth and development. In *Plants and the daylight spectrum*, H. Smith (ed.), 443–60. London: Academic Press.

Thomas, B. & D. Vince-Prue 1984. Juvenility photoperiodism and vernalisation. In *Advanced plant physiology*, M. B. Wilkins (ed.), 408–39. London: Pitman.

Tobin, E. M. & J. Silverthorne 1985. Light regulation of gene expression in higher plants. *Ann. Rev. Plant Physiol.* **36**, 569–93.

Vierstra, R. D. & K. L. Poff 1981a. Mechanism of specific inhibition of phototropism by phenylacetic acid in corn seedlings. *Plant Physiol.* **67**, 1011–15.

Vierstra, R. D. & K. L. Poff 1981b. Role of carotenoids in the phototropic response of corn seedlings. *Plant Physiol.* **68**, 798–801.

Vince-Prue, D. 1975. *Photoperiodism in plants*. Maidenhead: McGraw-Hill.

Vince-Prue, D. 1983a. Photomorphogenesis and flowering. In *Photomorphogenesis*, Encycl. Plant Physiol. NS 16B, W. Shropshire & H. Mohr (eds), 457–90. Berlin: Springer.

Vince-Prue, D. 1983b. The perception of light–dark transitions. *Phil Trans R. Soc.* **B303**, 523–36.

Vince-Prue, D. & A. E. Canham 1983. The horticultural significance of photomorphogenesis. In *Photomorphogenesis*, Encycl. Plant Physiol. NS 16B, W. Shropshire & H. Mohr (eds), 518–44. Berlin: Springer.

Virgin, H. I. & H. Egneus 1983. Control of plastid development in higher plants. In *Photomorphogenesis*, Encycl. Plant Physiol. NS 16A, W. Shropshire & H. Mohr (eds), 289–311. Berlin: Springer.

Wellman, E. 1983. UV radiation in photomorphogenesis. In *Photomorphogenesis*, Encycl. Plant Physiol. NS 16B, W. Shropshire & H. Mohr (eds), 745–56. Berlin: Springer.

Withrow, R. B. 1943. Radiant energy nomenclature. *Plant Physiol.* **18**, 476–87.

Zeiger, E. 1983. The biology of stomatal guard cells. *Ann. Rev. Plant Physiol.* **34**, 441–75.

Index